W9-CLE-332

DINOSAURS and Other Mesozoic Reptiles of CALIFORNIA

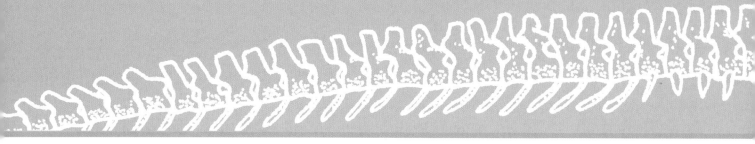

RICHARD P. HILTON

UNIVERSITY OF CALIFORNIA PRESS

Berkeley   Los Angeles   London

DINOSAURS

and Other Mesozoic Reptiles of

CALIFORNIA

ILLUSTRATED BY KEN KIRKLAND

FOREWORD BY KEVIN PADIAN

QE
861.8
.C2
H55
2003

University of California Press
Berkeley and Los Angeles, California

University of California Press, Ltd.
London, England

© 2003 by the Regents of the University of California

Library of Congress Cataloging-in-Publication Data

Hilton, Richard P.
     Dinosaurs and other Mesozoic reptiles of California /
Richard P. Hilton ; illustrated by Ken Kirkland ; foreword
by Kevin Padian.
          p.   cm.
     Includes bibliographical references and index.
     ISBN 0-520-23315-8 (cloth : alk. paper)
     1. Dinosaurs—California.   2. Reptiles, Fossil—California.
3. Paleontology—Mesozoic.   I. Title.
     QE861.8.C2 H55   2003
     567.9'09794—dc21   2002003613

Manufactured in Singapore
10   09   08   07   06   05   04   03   02   01
10   9   8   7   6   5   4   3   2   1

The paper used in this publication meets the minimum requirements
of ANSI/NISO Z39.48-1992 (R 1997) (*Permanence of Paper*).♾

This book is dedicated to my parents, Francis and Phyllis Hilton,

whose keen interest in natural history was the ultimate source of the inspiration for this book,

and especially to my wife, Kristin,

whose love, help, and support made it possible.

CONTENTS

ILLUSTRATIONS

FOREWORD

California is probably not the first state where one might expect to find dinosaur remains. Montana, Wyoming, Arizona, Utah, Colorado, South Dakota, New Mexico, Texas—even New Jersey, the first "dinosaur hunting grounds" in the United States—spring more readily to mind. But California, one of the largest states and one of the most complex geologically, contains its share of dinosaur fossils. It also boasts a great diversity of other fossil reptiles that lived at the same time as the dinosaurs, especially marine reptiles, which are among the best examples in the world.

Because much of California was under the sea during the age of dinosaurs, marine reptile fossils are relatively common. Land-dwelling reptiles such as dinosaurs, birds, and tortoises occasionally washed down rivers and streams, where they bloated, floated, and decayed, leaving their less common skeletons in the mud and sand of this marine environment as well. As our frequent earthquakes remind us, the state is geologically active. In addition to producing ground-shaking, this activity uplifts ancient bottom sediments into ranges of hills and mountains. Through time, those uplifted sediments have been eroded and fossils of reptiles exposed. Another relevant process is subduction, in which surface rocks are drawn back into the Earth as adjacent crustal plates grind into each other. Because the fossils are dragged along with the rocks, their source and geologic history often end up scrambled and difficult to understand.

In *Dinosaurs and Other Mesozoic Reptiles of California,* Dick Hilton has pieced together a fascinating history of Mesozoic fossil collecting in the state. Most of the first paleontological pioneers have now passed away, and their families, colleagues, friends, and memories are scattering, so this book is timely. As the Irish poet says, "Their like will never be again." Reptile fossils were collected in California by people who found tremendous joy and satisfaction in their discovery. Long before freeways and modern conveniences, these paleontological pioneers endured harsh weather, cold, rain, fires, and the disasters of field work to patiently track their quarry. And long before so many great fossil resources were destroyed by excavations, pavement, and housing developments, their efforts resulted in the great collections that we have in our museums today. Now their places are being taken by a new generation of amateur paleontologists—"amateurs" in the true sense: lovers of the pursuit.

Dick, who has spent countless weeks and months in the company of these dedicated fossil hunters, plays a crucial role in bridging the communities of scientist and amateur. His book presents not only information about the fossils and where they are found but also the stories of the people who are collecting, preparing, and preserving them. In addition, he has pored through libraries and museum archives, searched out field notes and personal reminiscences, and interviewed the remaining pioneers to weave their exploits with those of today's collectors. He has solicited help from fossil experts all over the country to provide authoritative identifications and information. This book can therefore be perused as a reference, read as a text, or enjoyed as a chronicle of the discoveries of Mesozoic fossil reptiles in California. It is a worthy companion to the best dinosaur books as well as to accounts of the state's geological past such as John McPhee's *Assembling California.* With this contribution, Dick Hilton provides an important piece of California's past that would otherwise be lost.

KEVIN PADIAN
Museum of Paleontology
University of California, Berkeley
April 2003

ACKNOWLEDGMENTS

I would like to express my sincere gratitude to numerous people and institutions that have helped in the formulation of this book.

First I would like to thank my colleagues at Sierra College. These include author and dinosaur paleontologist Frank DeCourten and biologists Charles Dailey, Shawna Martinez, Joe Medeiros, Ernie Riley, and Jim Wilson; photographer Rebecca Gregg; historian Lynn Medeiros; and Sierra College president Kevin Ramirez and vice presidents Fred McElroy, Morgan Lynn, and Ron Martinez.

Several libraries and their staff members made essential contributions. The Bancroft Library at the University of California, Berkeley, is where I conducted the bulk of my library research. I also frequented the library of the California Geological Survey in Sacramento, where I would like especially to thank information geologist Dale Stickney for his generous assistance. Cathy McNassor was extremely helpful in finding historical letters in the Chester Stock Library of the George C. Page Museum of the Natural History Museum of Los Angeles County, and she and librarian Don McNamee of the Natural History Museum provided me with many historical photographs; John Sullivan of the Huntington Library, San Marino, was instrumental in getting copies made. I obtained much information, including quotations, from the Archives of the Museum of Vertebrate Zoology at Berkeley and at the California Institute of Technology Archives in Pasadena, where head archivist Judy Goodstein was very helpful. In addition, I would like to thank the staffs of the California State Library in Sacramento, the Bioscience and Natural Resources Library at the University of California at Berkeley, the Physical Science and University Libraries at the University of California at Davis, and the Sierra College Library in Rocklin, California.

I visited many museums in the search for fossils and historical information. The University of California Museum of Paleontology in Berkeley provided a wealth of help and information; here I am grateful to collections managers Pat Holroyd and Karen Wetmore Grycewicz; preparator Jane Mason; paleontologists Nan Crystal Arens, Wyatt Durham, Diane Erwin, Mark Goodwin, Joseph Gregory, Howard Hutchison, Ryosuke Motani, James Parham, Howard Schorn, Thomas Stidham, and Samuel Welles; graphics assistant David Smith; and Judith

Scotchmoor, director of public programs for the museum. At the University of California, Berkeley, Museum of Vertebrate Zoology, I would like to thank museum scientist Barbara Stein. The Natural History Museum of Los Angeles County and its staff were most helpful, particularly collections manager Samuel McLeod; paleontologists Larry Barnes, Louella Saul, J. D. Stewart, and David Whistler; and preparator Gary Takeuchi. At the San Diego Natural History Museum, collections manager Fritz Clark and paleontologists Richard Cerruti, Thomas Deméré, and Bradford Riney provided valuable assistance. I would also like to thank paleontologist Robert Reynolds of the San Bernardino County Museum of Natural History; paleontology collections manager Jean DeMouthe and curatorial assistant Annette Fortin of the California Academy of Sciences; and Mary Ann Turner, collections manager of vertebrate paleontology at the Yale Peabody Museum. Acknowledgment is also due Lisa Babalonia and Jay Michalsky at the Ralph B. Clark Park Paleontological Museum in Buena Park.

The following scientists lent their expertise: Gorden Bell of the National Park Service; Michael Greenwald of the South Dakota School of Mines and Technology, Museum of Geology; Mel Bristow of Monterey Peninsula College; Kenneth Carpenter and Kirk Johnson of the Denver Museum of Natural History; Darrel Cowen of the University of Washington; Richard Cowan of the University of California, Davis; Philip Currie and Elizabeth Nicholls of the Royal Tyrrell Museum of Palaeontology, Drumheller, Alberta, Canada; Donald Dupras of the California Geological Survey; Gregory Erickson of Florida State University; John Horner of the Museum of the Rockies, Bozeman, Montana; Wann Langston of the University of Texas, Austin; Mervin Lovenburg of Modesto Junior College; David Mattison of Butte College, Oroville; Paula Noble of the University of Nevada, Reno; Robert Purdy of the National Museum of Natural History, Smithsonian Institution, Washington, D.C.; Timothy Rowe at the University of Texas; and Glenn Storrs of the Cincinnati Museum Center.

Many scientists and individuals associated with the discovery of California Mesozoic reptiles contributed greatly, including Patrick Antuzzi, Edwin Buffington, Gino Calvano, Gerard Case, Leon Case, Geoff Christe, Steven Conkling, Jon Cushing, Phillip Desatoff, Robert Drachuk, Patrick Embree, Steven Ervin, Harley Garbani, Eric Göhre, James Guyton, Barbara Hail, Robert Hansen, Katie Heiger, Paul Henshaw Jr., Loralee Hopkins, Robert Hopper, Robert Hoy, James Jensen Jr., James Jensen III, Inez Kelly, Mahlon Kirk, Malcolm Knock, Robert A. Long, Steven Lozano, Doug Maitia, Paul Mayo, Charles McDougall, Robert Merrill, Nancy Muleady-Mecham, Frank G. Newton, David Peters, Lloyd Pray, Allen Seagreave, Robert Sharp, June Skalecky, Justine Smith, Buck Tegowski, Dale Turner, Margaret Waegell, Donna Wahl-Hall, Ryan Warren, Robert Wilson, and Carl Zarconi.

Bill Nelson, Kendal Schinke, Joe Medeiros, and Matthew Peak provided graphic illustrations. Pilot Wayne Sutton helped obtain aerial photography. Howard Allen lent a hand finding historical information, and Jennifer Kane helped to locate somewhat obscure references. Special thanks go to David Maloney, George Bromm, and Patrick Embree for their meticulous preparation of some of the fossil reptilian material at Sierra College.

A few individuals important in the history of discovery of Mesozoic reptiles in California provided eyewitness accounts and other valuable evidence, without which this book would be greatly diminished. These pioneers include Allan Bennison, the discoverer of the first dinosaur in California and the mosasaur *Plotosaurus bennisoni*, who supplied photographs and first-hand information about early discoveries in the San Joaquin province; Arthur Drescher, field crew leader and fossil discoverer for Chester Stock in the Panoche Hills; Douglas Macdonald, a participant in one of William Morris's Baja expeditions; and Robert Wallace, an expert on the San Andreas Fault and one of Drescher's crew in the Panoche Hills. William Morris was helpful in clarifying information about fossil reptile discoveries in Baja California. William Otto was kind enough to answer questions about fossil preparation and exploration under Chester Stock at the California Institute of Technology. Mark Roeder introduced me to his crew in Orange County, who were responsible for important fossil reptile discoveries; he also took me to the sites of these discoveries. The Staebler family—Art, Chad, and Dianne Yang-Staebler—were invaluable in helping me sort out the long history of Staebler family discovery in the Panoche Hills area; they also provided interesting stories and photography.

Artist Ken Kirkland not only lent his talent in illustrating the book but also helped me better understand the extinct reptiles we are trying to bring to life.

I would like to thank all of the staff at the University of California Press who helped make this book possible. I would especially like to thank science editor Blake Edgar, project editor Mimi Kusch, developmental editor Eric Engles, copy editor Anne Canright, designer Victoria Kuskowki, and proofreader Annelise Zamula for their many, many hours of effort and helpful advice.

And without the help, support, and early editing of Kevin Padian of the University of California Museum of Paleontology this book may not have been published.

Above all I wish to thank my wife, Kristin, for her patience and countless hours of editing, and for helping maintain our sanity and sense of humor throughout the project.

PREFACE

In 1993 I was looking for ammonites—extinct relatives of the chambered nautilus—in the hills east of Redding, California, with geologist Tom Peltier and my son Jakob Hilton when I happened upon some fossil bones. Two years later, when my friend and fellow fossil hunter Pat Embree removed the bones from the rock, I realized their importance. The bones comprised the lower leg and foot of a hypsilophodontid, a relatively small plant-eating dinosaur: the first of its kind from California.

After publishing a short article on the find in *California Geology* magazine I received a letter from geologist Allan Bennison, who as a boy had found the first dinosaur remains in California. The remains belonged to a hadrosaur, one of the duck-billed dinosaurs. He invited me to visit the site of that 1936 discovery as well as nearby sites where he had found other important reptilian remains, including the complete skull of a mosasaur, a large seagoing reptile that was subsequently named in his honor *(Plotosaurus bennisoni)*.

During that trip, I had a growing sense that a story was waiting to be told. My hypsilophodontid and Bennison's hadrosaur and mosasaur were just a few of the many Mesozoic reptile fossils that had been discovered in California. We have bones from herbivorous and carnivorous dinosaurs, giant turtles, ichthyosaurs, plesiosaurs, mosasaurs, and even flying reptiles. I thought that many people, including paleontologists, must be curious about these creatures—what they looked like, how they behaved, where they lived. Who found the fossils, and who prepared them out of the rock? Who were the scientists who studied them and published their findings? And what about the artists who brought the fossils to life?

In late 1999, after I had begun work on this book, I happened to have dinner with several colleagues, including one of California's best-known geologists. The subject of dinosaurs in California came up. This well-known geologist commented that he thought "perhaps" a dinosaur find had been made somewhere in southern California; the discussion then turned to the occasional mammoth remains that are sometimes found in the northern part of the state.

This conversation confirmed my suspicion that the great majority of Californians, including even geologists and some paleontologists, are unaware of the wealth of Mesozoic reptilian remains that have been found—and that continue to be found—here. In just the five years I spent doing research for this book, several plesiosaur remains, a couple of mosasaur remains, three more dinosaur finds, new turtles, the first Mesozoic birds, and the first pterosaur remains were found in California. It is my hope, therefore, that this book will enlighten many people about the wealth of Mesozoic reptile finds made in this part of the world and perhaps spur interest that will lead to many new discoveries.

In addition to describing California's dinosaurs and other fossil reptiles, this book tells the stories of the people behind their discovery. Within this group one would expect to find an abundance of scientists with training in paleontology (the study of prehistoric life), and there are a few. Certainly without them many of these discoveries would never have been brought to science. But we also find ranchers, weekend fossil hunters, community college teachers, and plenty of students at all educational levels. People of different races and ages have participated in this fascinating quest, as have a fair number of women, from the very beginning of discovery nearly a hundred years ago. Back then, vertebrate paleontology was considered a man's science, and even at midcentury women were relatively rare in the field, so their involvement is especially notable.

Even today, anyone with an interest in paleontology and who works properly and with the right people can make important contributions. Although we live in a technological world, someone still has to venture out to the rock outcrops in search of the fossils—an aspect of the field that has not changed since the 1800s. I hope it never does. The thrill of discovery is as inspiring for the learned scholar as it is for any eight-year-old fascinated by dinosaurs.

A warning is in order here: *The collecting of fossil bones is illegal on public lands unless proper permits are first obtained. On private land, never trespass and please do not ask landowners to allow you on their land* unless you are a qualified paleontologist; even then it is imperative to get written permission. If a novice stumbles across vertebrate fossils on public lands, the fossils must be left alone and a scientist summoned to the site. The best way for novices to become legitimately involved in the search for vertebrate fossils is to hook up with a working paleontologist or a professor of paleontology.

Remember that fossils do not belong on the mantel as mere curios. Fossils are pieces of a giant puzzle that reveals the history of life on this planet. They need to be appropriately housed in a scientific institution so that they are available for systematic study. Often a specimen lurking in a drawer of a properly curated collection has provided the very clue needed to answer a vexing question. In contrast, most fossils that are brought home and put in a drawer eventually end up in the trash.

If you have information that I might consider adding to the next edition of this book, please send me an e-mail: rhilton@sierracollege.edu.

INTRODUCTION

ON JUNE 11, 1936, seventeen-year-old Allan Bennison pedaled his bicycle from his San Joaquin Valley home near Gustine to the hills of western Stanislaus County, thirty-five miles away. For two years—inspired by high school fossil-collecting trips when living in Monterey County—he had been scouring the stream-cut canyons where 70-million-year-old marine rocks were exposed, looking for fossil shells. On this day he found something unexpected: fossilized bones. Bennison reported his discovery to his high school science teacher, M. Merrill Thompson, and together they alerted the paleontology department at the University of California, Berkeley. Paleontologist Samuel P. Welles, curatorial assistant Curtis Hesse, and artist Owen J. Poe went to the site to investigate. The bones proved to be the vertebrae and leg bones of a duck-billed herbivorous dinosaur known as a hadrosaur. Bennison had found California's first dinosaur.

Bennison's important discovery is one of many in the history of California paleontology. Ever since 1893, when Stanford professor James Perrin Smith uncovered ichthyosaur fossils in the Klamath Mountains, paleontologists, geologists, and amateur bone hunters have been finding the remains of Mesozoic reptiles in the rocks of California—everything from huge, fishlike reptiles that swam through the ocean to flying reptiles with eighteen-foot wingspans, ancient turtles and crocodiles, and dinosaurs such as Bennison's hadrosaur. We now have enough evidence of California's Mesozoic past to make reasonable interpretations about the reptilian species that lived here and to reconstruct their environments.

This book is about the Mesozoic reptile fossil discoveries made in California during the last hundred-plus years. It describes the fossils and what they tell us about the animals they were a part of, and it chronicles the efforts of those who made the discoveries. Although the emphasis is on dinosaurs, all the reptile groups for which we have evidence are covered. Because

THE FIRST RECORD OF A DINOSAUR FROM THE WEST COAST

ALTHOUGH the Cretaceous deposits of California are extensive and many thousand invertebrate fossils have been collected from these rocks, vertebrates of any kind are exceedingly rare. A few sharks' teeth and fish scales have been collected in this series; but evidence of the reptilian life, so common elsewhere, has been totally lacking until the present time. Some weeks ago, Mr. Allan Bennison, an astute high-school student, found a vertebra in an exposure of the Moreno Cretaceous near Gustin, Calif. Mr. Bennison had been collecting invertebrates in this area for some time, and realized that this find was important. It was forwarded to Dr. G. Dallas Hanna, of the California Academy of Sciences, San Francisco, who, in turn, brought it to the Museum of Paleontology, University of California. It proved to be a pre-sacral vertebra of a *Phythonomorpha*, probably of the *Platycarpus-Tylosaurus* group.

Mr. Bennison continued his work in this region, not content to rest after having turned up the first recognizable reptile from the California Cretaceous. In June, in the same Moreno formation (Upper Cretaceous), near Patterson, Calif., Bennison discovered the first specimen of the dinosauria from the West Coast Cretaceous. The material is very fragmentary and seems to represent only the hind quarters of the animal. There are twenty-seven vertebrae (caudal), parts of the foot and the ends of some of the posterior limb elements. There are over 500 fragments of bone, from which, it is hoped, enough may be "pieced" together to make an accurate determination of the form represented. It is, of course, not possible to definitely determine the genus; but from the recognizable fragments found, it appears to be a member of the Hadrosauridae, a "duck-billed" or Trachodont-like form.

CURTIS J. HESSE
S. P. WELLES

FIGURE 1

"The first record of a dinosaur from the West Coast." Hesse and Welles 1936.

northern Baja California has yielded numerous fossil remains of dinosaurs that must have roamed in what is now the state of California, these discoveries are included as well.

The book is divided into two parts. In part 1, the first chapter describes the long reach of time during which the wonderful reptiles of the Mesozoic evolved as well as the tectonic and ecological settings in which these animals lived. Chapters 2 through 4 then paint a written and visual picture of every Mesozoic reptile that has been found and identified from the Californias. The dinosaurs, which so capture the imagination, come first. These range from the herbivorous hadrosaurs, ankylosaurs, and hypsilophodonts to carnivores such as the tyrannosaurids and ornithomimids. In chapters 3 and 4 we meet a cavalcade of other exciting reptilian creatures: the winged pterosaurs and the flying dinosaurs that today we call birds, and the fishlike ichthyosaurs, together with other marine reptiles such as thalattosaurs, plesiosaurs, mosasaurs, and turtles.

The last two chapters of the volume make up part 2. Here the human side of Mesozoic reptile paleontology in California is chronicled. This history of discovery, preparation, curation, and publishing of the Mesozoic reptiles found in California is a province-by-province journey starting in the Klamath Mountains in the north and then proceeding on to the Sierra Nevada, Great Valley, and Coast Range in the middle of the state. It concludes with astonishing discoveries made in southern California and Baja.

On these adventures you'll meet the teenager who found the first California dinosaur, the fireman who found the first theropod, the paleontologist who found most of the flying reptiles in the state, and a husband, wife, and father who teamed up to find more Cretaceous remains than anyone in California. You'll meet scientists, amateurs, and even a dog that dug up a fossil reptile bone. You'll experience an earthquake, snakes, bears, scorpions, and a hurricane. You'll ride horses and mules, trains and ferries, buggies, cars, and trucks, and hike uncountable miles of rugged terrain. This exciting story of discovery will inspire your imagination as it presents creatures literally out of this world and describes adventures in their discovery that are unique in time and circumstance.

A complete glossary and bibliography are provided, as well as a list of museums and websites one can visit to experience and learn more. In addition, to add to the science and interest of the book, a detailed table of all of the Mesozoic reptiles found in California and Baja California is included in the appendix.

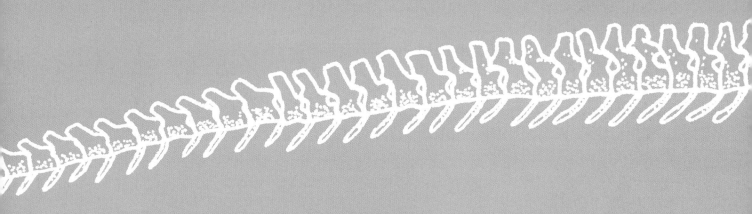

PART I

CALIFORNIA DURING THE AGE OF REPTILES

SPANNING 180 MILLION YEARS, the Mesozoic Era, which began 245 million years before present (MYBP) and ended only 65 MYBP, is relatively recent in the 4.5-billion-year history of Earth. Fossils of primitive life found in rocks more than 3.5 billion years old show us that life has been evolving on Earth for an extremely long time. By the middle of the Mesozoic, life had become very sophisticated. This was especially true of reptiles, which had radiated out to fill a myriad of niches in a wonderful variety of forms.

On land, dinosaurs were especially successful, and each dinosaur evolved to fit a particular way of life. We can get some idea of how highly evolved they were by looking at their

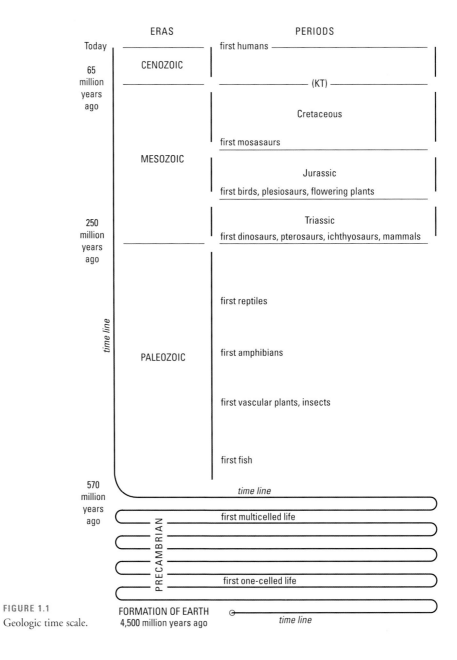

ERAS   PERIODS

Today

CENOZOIC

65 million years ago

first humans

——— (KT) ———

Cretaceous

first mosasaurs

MESOZOIC

Jurassic

first birds, plesiosaurs, flowering plants

250 million years ago

Triassic

first dinosaurs, pterosaurs, ichthyosaurs, mammals

first reptiles

time line

PALEOZOIC

first amphibians

first vascular plants, insects

first fish

570 million years ago

time line

first multicelled life

PRECAMBRIAN

first one-celled life

FIGURE 1.1
Geologic time scale.

FORMATION OF EARTH
4,500 million years ago   *time line*

skeletons and teeth. Some were slow, lumbering, plant-eating animals, while others were swift, sleek predators. Ichthyosaurs, a group of marine reptiles, left the land and evolved streamlined, fishlike skeletons, while pterosaurs and birds lightened their bones and modified their forelimbs to take to the sky. Evidence indicates that birds, wonderfully diverse creatures that dominate the skies and dazzle us with their grace and beauty, evolved from dinosaurs.

Many other significant evolutionary events occurred during the Mesozoic, among both animals and plants. Vascular plants had colonized continental interiors in the Paleozoic (the era that precedes the Mesozoic, extending from 550 MYBP to 245 MYBP), at which point some animals were able to move inland. By the Mesozoic, both terrestrial animals and forests had existed for a long time. An early Mesozoic forest was lush, with varieties of ferns, cycads, ginkgos, and conifers. As the era progressed, flowering plants evolved, becoming a significant food source for herbivorous dinosaurs, birds, mammals, and insects. The Mesozoic is often considered the age of reptiles, but most of mammalian evolution occurred then too, and amphibians born of the Paleozoic made a leap forward in the form of frogs then as well. Insects, too, continued their success story, one that also began in the Paleozoic, about 200 million years before the dinosaurs.

In the marine environment, the evolution of single-celled life forms continued apace. Their skeletons are important for dating some of the Mesozoic rocks of California, in cases where few or no larger fossils exist. Marine invertebrates also did well in the Mesozoic. Modern corals first appeared during that era, and echinoderms (such as urchins and starfish) became much more abundant. Mollusks, the animals that make seashells, evolved in a variety of forms, including the shelled cephalopods (the relatives of squid and octopus). Ammonites, a group of specialized cephalopods, are important fossils for dating many of the Mesozoic marine rocks of California.

The Mesozoic was a glorious time for life on Earth, but evidence suggests it ended with a bang 65 million years ago. In a great catastrophe, about 80 percent of all species died when an asteroid or comet hit the Earth, severely disrupting the web of life. Among the victims were the dinosaurs, who vanished, leaving only their fossils as traces of their long history of dominating the land.

1

GEOLOGIC HISTORY

AT THE BEGINNING of the Mesozoic Era, much of what is now California consisted of islands and ocean bottom. By the end of the Mesozoic, California had grown considerably, and many of its major features—the Klamath Mountains, the Sierra Nevada, the Mojave Desert, and the Peninsular Ranges—were present in their early forms. These geologic changes, which corresponded to changes in the environments in which dinosaurs and other reptiles could live, were the result of large-scale movements in the Earth's crust, driven by the process of plate tectonics.

PLATE TECTONICS

The crust of the Earth resembles the broken shell of a hard-boiled egg clinging to the white of the egg below. Each section of crust is called a plate, of which there are about a dozen. A plate may be composed of oceanic crust (made primarily of the black lava called "basalt") or oceanic and continental crust combined (continental crust being composed of many types of rock but with the average chemical composition of granite). Unlike the pieces of eggshell, however, each plate on the Earth's surface is moving, and has been for millions of years. The process that drives this motion, plate tectonics, involves two basic activities: the production of oceanic crust at spreading ridges; and the recycling of oceanic crust back into the Earth's interior, with the building of continental crust as a by-product.

The crustal plates of the Earth move because the mantle—the hot, iron- and magnesium-rich rock below the crust—is in slow but constant motion; the overlying crust simply goes along for the ride. Although the mantle is mostly solid, it slowly deforms, much like glacial ice. But whereas glacial ice can move at the lightning speed of a foot or more in a

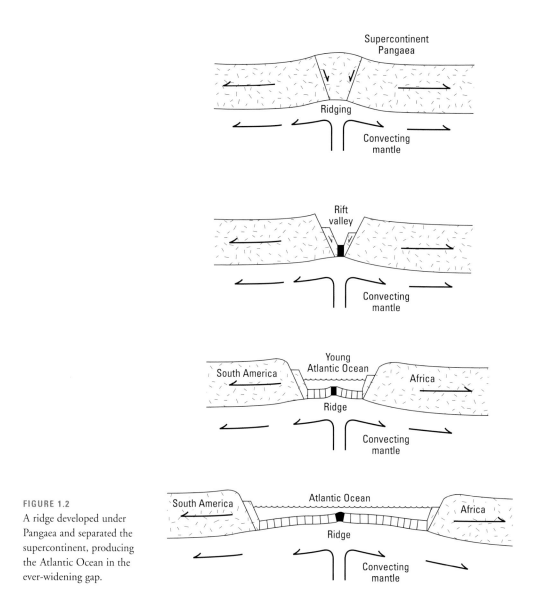

FIGURE 1.2
A ridge developed under Pangaea and separated the supercontinent, producing the Atlantic Ocean in the ever-widening gap.

single day, the mantle moves a mere inch or two a year—about the same rate that fingernails grow. As sluggish as this may seem, an inch a year multiplied by 100 million years adds up: you can see how continents as well as ocean crusts can move great distances over long periods of time. In fact, since the Mesozoic, when the Americas and the Old World began drifting away from each other, the Atlantic Ocean has opened up and spread about three thousand miles.

Where plates diverge, ridges well up, forming spreading boundaries. The mantle, driven by the heat below, rises and spreads out just beneath the crust at the ridge. Because of the

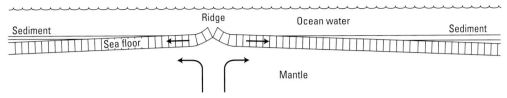

FIGURE 1.3

As crust moves away from the ridge, sediment accumulation increases on the older and older crust.

tensional spreading at the ridge, the pressure is a bit lower, allowing the mantle to become molten. As the mantle diverges, the brittle oceanic crust begins to stretch. After many decades of stretching, eventually the crust will crack all the way to the molten mantle several miles below. Because the basaltic ocean crust is relatively inelastic, it can't stretch far, so the crack that is produced is only a few feet wide—but it may be tens of miles long and miles deep. Under the pressure caused by the weight of the ocean crust above, magma from the molten mantle now gets injected into the fissure and spills out on the seafloor. (Imagine a waterbed filled with hot molasses; now imagine taking a razor blade and slicing through the thick plastic of the mattress.) Because the crack is thin and surrounded by relatively cool rock, the magma chills and crystallizes quickly. Any magma that spills out on the surface is further chilled by the cold ocean water. In the crack the congealed magma (sheeted dike) becomes the newest vertical slice of oceanic crust.

So it is at the ridges that over time the new crust of the Earth continuously forms. Every hundred years or so a new crack opens and is filled with magma separating the previous flow of basalt off to the sides. Crack and fill by crack and fill, the ocean floor expands, moving in opposite directions from the ridge. As its distance from the ridge increases, the crust becomes the repository of more and more sediment (usually mud). Over millions of years, the debris stacks up, forming sedimentary layers thousands of feet thick (see fig. 1.3). Eventually the oceanic crust may run into another plate—a continent or another piece of oceanic crust— that is either stationary or moving in a different direction. It is here that the muds and sands are scraped off and added to islands or continents.

TECTONIC PROCESSES IN MESOZOIC CALIFORNIA

The Mesozoic plate tectonic settings described here for ancestral Mesozoic California not only illustrate a history of the tectonics of this part of western North America but also provide

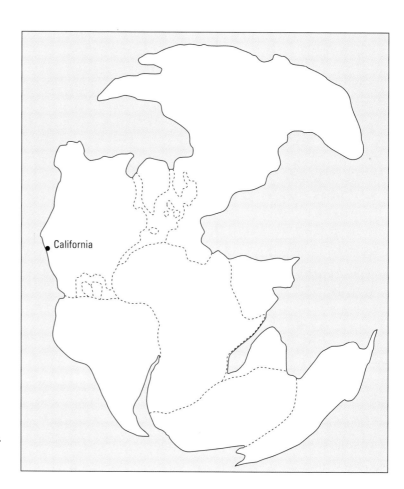

California

FIGURE 1.4
During the Triassic,
Pangaea comprised all
of the present continents.
Modified from Drewry
et al. 1974.

depositional settings in which Mesozoic reptiles were fossilized. In what follows, I cover the three periods of the Mesozoic—the Triassic, Jurassic, and Cretaceous—in some detail to provide a grounding for the animal life of this critical geologic era.

The Late Triassic

In the Late Triassic (ca. 235 MYBP) the western edge of what is now North America was growing westward by the accretionary processes of plate tectonics. At this time, a portion of the ancestral Pacific Ocean floor was spreading eastward away from a ridge and plowing into the continent. In a process called subduction, the seafloor, because it was lower than the continental plate and made of dense basalt, was forced under the neighboring plate, to be recycled into the mantle. This ancestral continent—called Pangaea by Alfred Wegener, a German meteorologist—stretched from California across what is now North America and continued on uninterrupted (except by the occasional shallow sea) through Europe and Asia

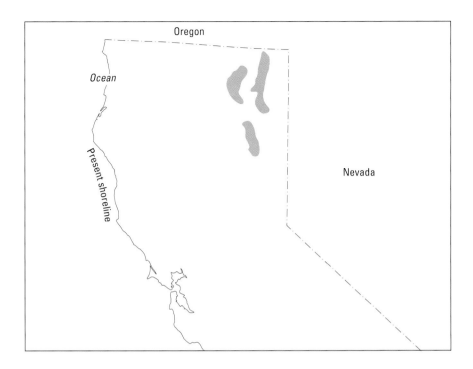

FIGURE 1.5
Representative islands of
northern California in the
Late Triassic.

to the east coast of what is now China. There was no Atlantic Ocean, just the one huge su-percontinent, which had been assembled by the coming together of previous continents im-mediately prior to the Mesozoic.

In the Late Triassic, northern California existed merely as a series of islands far from the coastline of Pangaea. Today the only evidence for these islands is found in limestones de-rived from ancient reefs and rounded gravels and sands originally from beach or stream de-posits and now found in sandstones and conglomerates (McMath 1966). The islands most likely possessed no large terrestrial environments in which evolving dinosaurs might have been preserved. The seas were warm and rich with reefs, and it was in these reefs' limy de-bris that reptile skeletons and associated remains were entombed.

The area that is today California and northwestern Nevada had been added to North America during events that occurred at the close of the Paleozoic era and beginning of the Triassic period. These events welded island arc materials (muds, sands, and lava, similar to the present-day Aleutian Islands) to the edge of the continental shelf at the western edge of North America (i.e., Pangaea) that stretched from southern California diagonally through central Nevada (see fig. 1.6). The new edge of the continent was in what is today's Eastern Klamath Mountains Province and the eastern portion of the northern Sierra Nevada (see fig. 1.7). The Late Triassic rocks of Shasta County were deposited on this Paleozoic addition to

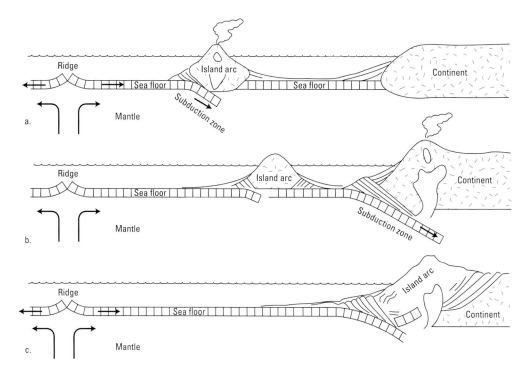

FIGURE 1.6
The approach and accretion of island arc materials to Paleozoic western
North America. (a) Island arc develops. (b) Island arc approaches North
America. (c) Island arc(s) accrete to North American continent.
Modified from Hannah and Moores 1986.

western North America. These rocks, the Hosselkus Limestone, hold the only Triassic reptilian fossils found in California.

In the Late Triassic, after the Hosselkus reefal limestones were laid down on the western
edge of this accreted arc material (see fig. 1.8), the complexity of plate tectonic activities becomes rather nightmarish. Several types of plate boundaries were now produced, involving
island arcs, rifting, San Andreas–type faulting, and the welding onto the continent of largely
oceanic sedimentary materials by converging plates. These activities led to the further accretion of arc materials, oceanic crustal materials, and both shallow- and deepwater ocean
sedimentary deposits.

Although the geologic changes of the Late Triassic were complex, they can be summa-
rized as follows: The seafloor moved toward and was subducted under the edge of the continent, in much the same way that the eastern Pacific seafloor is today diving under South

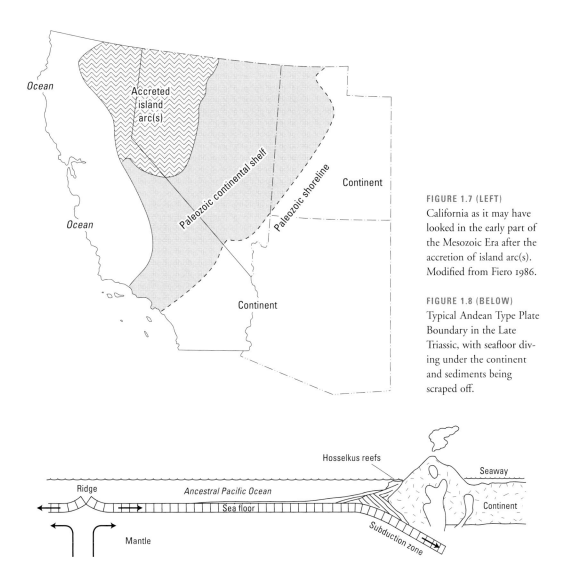

FIGURE 1.7 (LEFT)
California as it may have
looked in the early part of
the Mesozoic Era after the
accretion of island arc(s).
Modified from Fiero 1986.

FIGURE 1.8 (BELOW)
Typical Andean Type Plate
Boundary in the Late
Triassic, with seafloor diving under the continent
and sediments being
scraped off.

America. Here the subducting oceanic basalt was recycled into the mantle, but the seafloor sediments, being less dense, were scraped off and uplifted into mountains. This type of plate boundary is called an Andean Type Plate Boundary, after the Andes Mountains.

The Jurassic

In the middle of the Jurassic Period (ca. 160 MYBP) a dramatic change occurred. After intense mountain building in the ancestral Sierra, the western edge of the continent in the

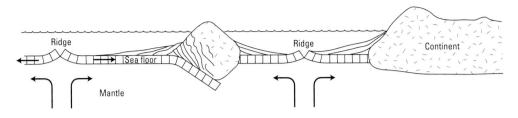

FIGURE 1.9
Rifting of the western edge of the Klamath-Sierra region during
the mid-Jurassic produced new oceanic crust. Modified from
Saleeby et al. 1994.

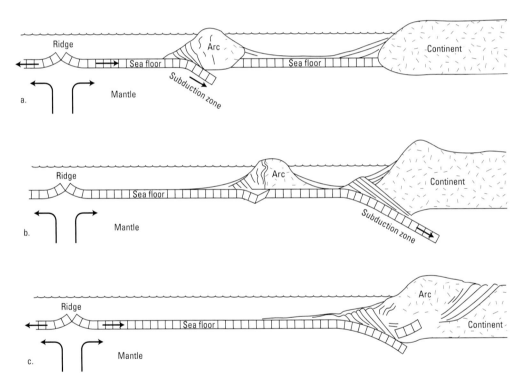

FIGURE 1.10
As seaward subduction on the offshore arc stopped and a new subduction
zone formed on the edge of the continent, the offshore arc merged with
western North America. (a) Near-shore rifting ceases. (b) Offshore arc
subduction ceases and a new subduction zone develops under the continent. (c) The island arc materials as well as ocean floor and ocean floor
sediments are accreted to the edge of western North America. Modified
from Saleeby et al. 1994.

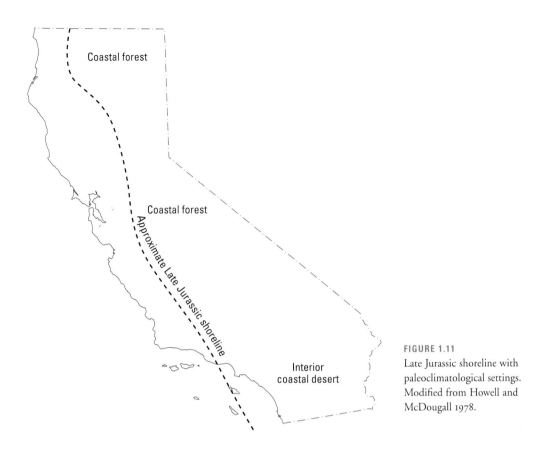

Coastal forest

Coastal forest

Approximate Late Jurassic shoreline

Interior
coastal desert

FIGURE 1.11

Late Jurassic shoreline with
paleoclimatological settings.
Modified from Howell and
McDougall 1978.

Klamath-Sierra region began to undergo ridgelike rifting (see fig. 1.9). Although similar to
that which took place at oceanic ridges, this rifting occurred under the edge of the continent
itself. As the edge of the continent began to rift away, new oceanic crust was produced in
the area between. During the Late Jurassic the rifting ceased and the subduction terminated
on the western side of the offshore island arc. A new subduction zone developed on the eastern side of the new ocean floor, up against the continent. This eventually caused the offshore
island arc, the seafloor sediment, and fragments of the seafloor itself to be jammed up against
(and to a certain extent thrust over) the continental edge, adding new girth to the continent
(see fig. 1.10). It was in this tectonic turmoil that many California Mesozoic reptile remains
were entrapped in sediments and preserved.

By the end of the Jurassic, enough material had been accreted onto California, and granitic
material intruded into the ancestral coastal Sierra Nevada, that the shoreline had migrated
west, probably extending through what is now the middle of the Sacramento and San Joaquin

FIGURE 1.12
Hogback, held up by well-
cemented Late Jurassic conglomerates, in the north-
western portion of the Great
Valley Province. Photo by
the author.

Valleys. A rugged Andean-type mountain range was forming west of the present border of the Sierra Nevada and the Sacramento Valley, and it was being intruded by granitic magmas. Today granitic rocks dip under the valley and extend for several miles to the west of the present western edge of the Sierra foothills. The overlying mountains had to have been many thousands of feet thick for these granitic rocks to form.

At about this time the central Mojave Desert was an interior coastal desert behind the western range of mountains and adjacent to an interior seaway. Evidence for this comes from the Jurassic Aztec Sandstone, which contains lithified cross-bedded dune sand.

Today in the hills on the west side of the Sacramento Valley we find ridges (hogbacks), some held up by well-cemented gravels of latest Jurassic age. These are the oldest beds of what is known as the Great Valley Group, the bottommost of several miles of latest Jurassic to latest Cretaceous beds (see fig. 1.12).

During most of this time the area of deposition of the Great Valley Group was a forearc basin between the ancestral Sierra and the trench to the west that actively accumulated sediment eroded off the Sierra, though some of the older beds of this sequence may have rafted in with an island arc. For nearly a hundred million years, layer upon layer of sediment was deposited here, entrapping skeletal remains of plesiosaurs, ichthyosaurs, mosasaurs, and turtles. The occasional dinosaur carcass would also wash into the sea and be deposited in these same sediments.

The Cretaceous

By the Late Cretaceous, the western edge of the Sierra Nevada had been eroded back and the sea was flooding into the areas of what are now the low Sierran foothills. From the Sierra Nevada and southern California's Santa Ana Mountain foothills all the way to the western Baja Peninsula, the forearc basin was accumulating sediment from the mountains to the east. To the west of the basin at the leading edge of the continent lay the trench. Sediments from the seafloor as well as other rocks transported by the seafloor crust were being swept up and accreted to the continent as the seafloor crust subducted beneath it (see fig. 1.13).

In the Cretaceous, this forearc basin accumulated layer upon layer of fine siltstones, shales, and occasional limestones. The sedimentary particles were very small, ranging from mud to silt, because the environments in which they were deposited were relatively still and unenergetic.

Punctuating this stillness, however, were avalanches of sediment carried by undersea debris flows known as turbidity currents. These turbidity currents, traveling at highway speeds, cascaded from nearshore environments, where sand and sometimes gravel would build up at river mouths and along beaches. As turbidity currents slow down, their larger, heavier particles generally settle out first, followed by the smaller grains of sand, silt, and clay. The resulting layers of thick, hard rock are called turbidites. The sediment that makes up a turbidite is often graded with coarse sediment toward the bottom and finer material toward the top. When the original material is nearly pure sand, however, the resulting turbidite may be a layer of sandstone.

Many of the Jurassic and Cretaceous reptilian fossils were deposited in the basin that is now the Great Valley Province, while many of the Late Cretaceous fossils are found in a similar tectonic setting in the Peninsular Ranges Province as well.

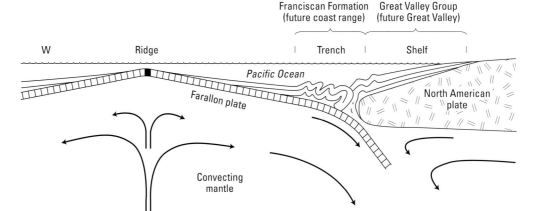

FIGURE 1.13

Late Cretaceous plate tectonic setting for central California. Note the Great
Valley Group deposited in the forearc basin, and the Franciscan Formation
formed at the trench, where continental and seafloor materials were
scraped off and added to the continent. Modified from Dickinson 1976.

MARINE DEPOSITIONAL SETTINGS

The tectonic activities that built California also provided marine depositional settings in which
fossils were preserved. The Late Triassic Hosselkus Limestone in the Klamath Mountains
Province in Shasta County has yielded numerous early marine reptiles. We know from microfossils and invertebrates found in these reefal limestones that the seas were warm and fer-
tile. Imagine a scene much like Australia's Great Barrier Reef, rich with fish and coral—but
instead of seals and dolphins, picture thalattosaurs climbing out on the reef and ichthyosaurs
chasing the fish (see fig. 5.2).

The oldest Jurassic marine rocks in California found to contain marine reptile fossils are
those of the Salt Springs Slate in the central Sierra foothills. The Salt Springs Slate is a probable equivalent to the Mariposa Formation, the rocks of which are Middle and early Late
Jurassic in age and were deposited either in a submarine fan sequence (Tuminas and Moores
1982; Tuminas 1983) or as trench fill between the continent and an approaching island arc
(Bogen 1984). Fossils of ammonites, clams, brachiopods, and even sea urchins are occasionally found in sedimentary rocks of the Mariposa Formation. The only plesiosaur (a marine
reptile) remains found in the Sierra Nevada Province were found in this formation. At the
time of the deposition of the Mariposa Formation the shoreline of central California was
somewhere in what is now the high Sierra or eastward.

In Orange County in southern California, vertebrae from an elasmosaurid (long-necked)
plesiosaur were found in the Jurassic Bedford Canyon Formation. These are the oldest fos-

FIGURE 1.14
Belemnite from the Great Valley Group. Photo by the author.

sil marine reptile remains from the southern half of the state. The Bedford Canyon Formation, like the Mariposa Formation, is composed of older bedrock that was involved in the mountain-building events that formed the ancestral Sierran and Peninsular Ranges.

The Great Valley Group found along the western edge of the Sacramento and San Joaquin Valleys contains many vertical miles of sedimentary rocks originally deposited along the Late Jurassic to Late Cretaceous continental edge. The oldest of these layers are latest Jurassic, and the more resistant of these are made of cemented gravels called conglomerate. It seems that turbidity currents, made of what were probably beach gravels, swept up the occasional bone that was lying on the muddy sea bottom. The bones were deposited with the gravel when the turbidity current slowed and came to rest.

Evidence of the turbidity currents comes with the presence in turbidites of rip-up clasts, broken pieces of belemnites, and occasional waterlogged tree trunks. Rip-up clasts are pieces of the muddy bottom that were pulled up by the fast-moving turbidity current and then carried to the site of deposition. Belemnites are the heavy internal skeletons of squidlike animals shaped like oversized ballpoint pens (see fig. 1.14). These, too, were probably sitting on the bottom and became caught up in the flow. The wood, now petrified, was most likely driftwood that had floated down rivers into the sea, became waterlogged, and settled to the bottom. Bones of plesiosaurs and even a dinosaur have been found in these turbidites as well.

Sequences of fine-grained layers punctuated by occasional sandstones are common in most of the Great Valley Group in the linear hills along the western edge of the Sacramento and San Joaquin Valleys. These layered sequences are called flysch (see fig. 1.15). Most reptilian fossils are found in the fine-grained portions of the flysches of the Great Valley Group, often as groups of bones or even articulated skeletons. Only here could a complete carcass settle to the bottom and remain relatively undisturbed while being slowly entombed by the fine

FIGURE 1.15
Typical flysch found in the
Great Valley Group. Photo
by the author.

sediment settling over it. Turbidites, in contrast, rarely yield more than a single bone or bone fragment, and complete skeletons are exceedingly uncommon.

Most Jurassic and Cretaceous reptilian fossils were deposited in relatively shallow water along the coast. An exception is two ichthyosaur fossils found in rocks originating in the Franciscan Formation of coastal California. Until the acceptance of the theory of plate tectonics in the early 1960s, the origin of this formation was a mystery. Even today geologists debate the origin of some of its rocks, though it is generally agreed that many rocks of the Late Jurassic to Early Cenozoic Franciscan Formation were originally deposited in the deep ocean far from North America. One such rock type is radiolarian chert. The two ichthyosaur fragments were found in radiolarian chert cobbles that originated in the Franciscan Formation.

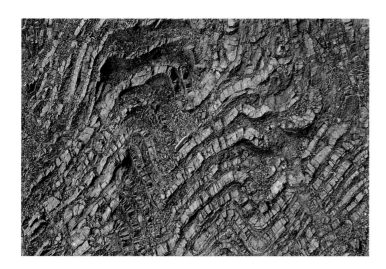

FIGURE 1.16
Radiolarian chert on the
Marin Headlands. Photo by
the author.

In the road cuts of the Marin Headlands at the north end of the Golden Gate Bridge one can see exceedingly fine outcrops of highly contorted, reddish-brown radiolarian chert (see fig. 1.16). These well-defined layers are an inch or more thick and made mostly of silicon dioxide, the material that makes up the mineral quartz. Chert is usually stained red or occasionally green, the result of minor amounts of iron in the rock combining with oxygen (rust). The quartz material in the chert is made of uncountable numbers of skeletons of microscopic creatures called radiolaria, one-celled microbes that live in the sea and feed on plankton (see fig. 1.17). Their beautiful skeletons are made from dissolved silicon dioxide absorbed from the seawater. When the radiolaria die they slowly settle to the bottom and accumulate in layers of soft ooze.

Radiolarian chert forms well offshore from any landmass where there is virtually no mud from rivers or streams to settle on the ocean bottom. Here the predominant sediment is made from the radiolaria skeletons themselves. Minor sources of mud do occur, however: wind-blown ash from volcanoes, dust blown off the deserts of the world, and meteoric dust from space. This dust slowly settles to the surface of the ocean, and eventually, and even more slowly, to the ocean bottom. It is this dust that provides much of the iron that oxidizes to produce the red and green colors of radiolarian chert, sometimes brilliant enough that rock-hounders call it jasper.

The chert was at one time lying as flat layers of radiolarian ooze on the open ocean floor. Eventually, as the seafloor and the continent converged, these layers were scraped up, folded, faulted, compressed, and cemented in the trench area of the subduction zone. Today scientists

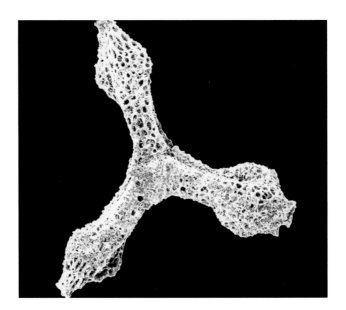

FIGURE 1.17
Jurassic radiolarian skeleton from the Franciscan Formation. Photo by Paula Noble.

etch the chert with acid to reveal the elegant radiolaria skeletons within. Because other identifiable fossils may be almost nonexistent in this rock, fossil radiolaria, distinctive in appearance and having evolved over time, sometimes provide the only way to date older rocks of portions of the Coast Range and other western mountains.

EVIDENCE OF MESOZOIC TERRESTRIAL ENVIRONMENTS

Because most of California's Mesozoic terrain was mountainous, it lacks the rich terrestrial depositional settings found in the continental interior. Nevertheless, some fossil reptile remains have been preserved in terrestrial sedimentary environments ranging from desert sands to mountain stream channels. Given that much of California was underwater then, we must also rely on marine depositional environments for evidence of most of California's terrestrial realm.

The Triassic

Almost nothing is known of California's Triassic terrestrial environment. We are fairly sure that islands rose above the sea in the northeast part of the state because we have reefal limestones of the Late Triassic age, some with corals that most likely formed near the shores of islands. Also present are minor conglomerates and rounded quartz sands, indicating the presence of beach, stream, or river environments. What we do not have is evidence of plant or terrestrial animal life from these islands. Were these islands barren deserts or lush and crawling with reptiles, amphibians, and insects? We may never know.

The Late Jurassic

In the Late Jurassic, what is now eastern California was land, while most of the western portion of the state still lay beneath the sea. A rich, lush forest covered the northern part of the state, complete with cone-bearing trees (including redwoods), leafy ginkgos, and an understory of ferns and cycads. It must have looked similar to a coastal redwood forest of today, but with moister summers and warmer winters it probably felt more like a jungle. It was a wonderful green steamy abode for life, with dinosaurs lurking in the dense foliage.

Because most of the rocks of this age are folded and faulted and many were heated and pressed, most fossil evidence has long been squeezed out of existence. Then too, steep mountains and a lack of a coastal plain running the length of the Jurassic west coast make it improbable that dinosaurs would be preserved in the first place. Despite these odds, the Trail Formation in the extreme northern Sierra Nevada has yielded a hint of Late Jurassic life. The Trail Formation was deposited in a terrestrial setting and is composed of braided stream deposits including conglomerates, sandstones, and shales. Fragmentary remains of dinosaurs are found here, the oldest dinosaur bones to be discovered in the state (Christe and Hilton 2001). Although the evidence is scant, just enough fossil-rich rocks have survived to give us hope that we might learn more.

In the southern part of the state at about this time the central Mojave was an interior coastal desert. Dinosaurs left their tracks in Late Jurassic windblown dune sands that later formed the Aztec Sandstone. These are the only Mesozoic reptile tracks from the Californias. A possible comparable setting is found today on the southwest coast of Africa, where elephants and antelopes wander dunes in search of the occasional group of plants or the seasonal stream channel lined with vegetation. Hence the Late Jurassic climate of California seems to have been somewhat similar to the climate today, with lush green forests in the north and a sandy desert in the southern interior.

In the sea during the latest Jurassic through the latest Cretaceous, for nearly a hundred million years layer upon layer of sediment was deposited on what was then the continental shelf. In central California, these layers were to become the Great Valley Group. Entrapped in them are plant fossils and the skeletal remains of many reptiles. The fossil plants found here hint that the lush mountain forests to the east persisted from the Late Jurassic to the close of the Mesozoic.

The Early Cretaceous

Sedimentary rocks of Early Cretaceous age can be found in many places from about Santa Barbara northward along the western side of the state. All of these rocks—mostly flysches containing fine-grained sediments, permitting good fossil preservation—were originally laid down in the sea. The shore at the time must have been about where the bases of interior mountains lie today (for the central part of the state, for example, at the base of the Sierra Nevada). Streams and rivers drained the interior, carrying wood, leaves, cones, seeds, and even carcasses out to sea. These eventually settled to the bottom to give us a partial record of Early Cretaceous terrestrial life.

This record indicates that the northern part of the state continued to be a forested environment similar in appearance to that of the late Jurassic. The partial remains of an herbivorous dinosaur, a hypsilophodontid, indicate that dinosaurs continued to roam the Early Cretaceous northern forest. Fragmentary remains of a pterosaur tell us that flying reptiles were here too. A lack of Early Cretaceous rocks in the southern part of the state makes it difficult to know what interior southern California was like during this time.

The Late Cretaceous

Evidence of the Late Cretaceous terrestrial environment in California is relatively rich. From Oregon to central Baja sedimentary rocks originally deposited as sediment in the sea contain traces from the land. In Baja California, too, rare Late Cretaceous terrestrial rocks yield important clues.

In northern California, most of our evidence comes from the Chico Formation, found along the eastern edge of the Great Valley. The forests apparently persisted, for we continue to see the seeds, cones, and branches of trees similar to the monkey puzzle (*Araucaria*) and redwood as well as the fronds and stems of ferns and cycads. For the first time we now have fossil leaves and seeds from broad-leafed flowering trees—and with flowers come fruits and more food for an increasingly diverse fauna. We find the shells of terrestrial snails that lived on these plants, and even the remains of a meat-eating dinosaur. Pterosaurs and birds filled these Late Cretaceous skies, and from rocks on the western side of the Great Valley we find a possible terrestrial turtle and several hadrosaur remains.

In the Peninsular Range, the mountainous province that begins at Los Angeles and extends over nine hundred miles southward to the tip of Baja California, are the remains of many more dinosaurs. In San Diego County we have hadrosaurian remains and an ankylosaur skeleton. In deltaic sedimentary rocks in the coastal Peninsular Range of Baja are fossil bones and teeth of several meat-eating dinosaurs as well as more hadrosaurs and other plant-eating dinosaurs. Crocodilians, lizards, birds, and even early mammals have been found as well.

Evidence of plants—the leaves of vines, ginkgo, and other deciduous trees—has been discovered in Baja as well, suggesting an environment similar to that of Late Cretaceous Alta California. Some fossil conifers (including *Araucaria* and redwoods) have been found in life position, with their roots still penetrating the sandstones below. Even the remains of palm logs have been found. The topography of the west coast of Baja was probably much the same back then as now, but instead of the harsh coastal desert of today, a rich green forest complete with reptiles, mammals, and birds made it a much more productive environment during the Late Cretaceous.

So from Shasta County in the north to central Baja California in the south a treasure trove of fossils gives us a glimpse through the keyhole of time: to a time much different than today, when dinosaurs roamed California's warm, moist forests and dry, hostile deserts.

2

THE DINOSAURS

THE TERM *DINOSAUR,* meaning "fearfully great reptile," was first used by the British anatomist Richard Owen in 1842, who recognized that fragmentary fossils that had been found locally in England were those of unusually huge reptiles. As more and more dinosaur fossils were found throughout the world it became apparent that dinosaurs were different from their other reptilian relatives, and eventually two major groups were distinguished.

Dinosaurs were the result of evolutionary changes that took place in reptiles during the Paleozoic Era. Approximately 325 million years ago, reptiles evolved from a basal tetrapod (ancestral four-legged vertebrate). By the close of the Paleozoic, about 250 million years ago, reptiles had become evolutionarily quite sophisticated, and a group of reptiles called diapsids had evolved (the term *diapsid* refers to two holes in the skull behind the eye socket). At the beginning of the Mesozoic, the dinosaurs arose from one branch of the diapsid stock, the archosaurs.

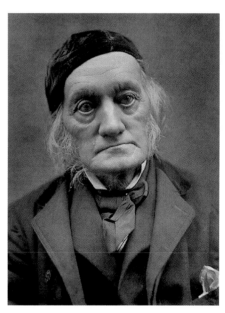

FIGURE 2.1
Richard Owen, about 1858. From the collection of Kevin Padian.

DINOSAUR CHARACTERISTICS

Most people think they know what a dinosaur is, but what are we talking about when we use the term? Dinosaurs have become real to us in books and movies, and many children can rattle off their names and tell you something about them. The trouble is, some people erroneously think that plesiosaurs (a group of sea-going reptiles) and pterosaurs (a group of flying reptiles) are dinosaurs too.

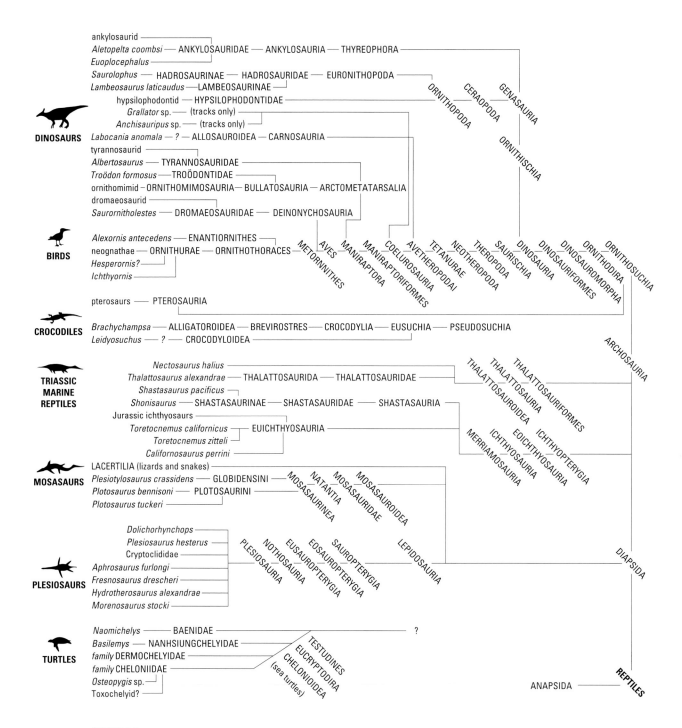

FIGURE 2.2

A "family tree" of California Mesozoic reptiles. After Bell 1997; Brochu 1999; Caldwell 1997; Chiappe 1997; Currie and Padian 1997; Motani 1999; Nicholls 1999; and Parham, pers. comm., 2000.

In fact, the term *dinosaur* is nothing more than a catch-all word for some specialized reptiles with upright postures (as opposed to the more squat lizards and crocodilians) that existed during the Mesozoic Era. Because no dinosaurs evolved to fit marine niches, the reptiles that lived exclusively in the sea (ichthyosaurs, plesiosaurs, mosasaurs, and turtles) were not dinosaurs. Neither were lizards, snakes, or crocodilians dinosaurs, even though they lived in some of the same places at the same time that dinosaurs did. And the flying reptiles (pterosaurs) were definitely not dinosaurs, though many scientists consider birds to be a specialized group of dinosaurs.

We normally consider dinosaurs to have been terrestrial animals, although some occasionally swam, and some may have led a fairly mooselike existence, spending time up to their bellies in water. Unlike most other land-dwelling reptiles, such as the squat-postured lizards and crocodiles, dinosaurs and their closest relatives evolved hip and leg bones that forced an upright walking or running position. Many were fast runners, and they didn't often drag their tails and bellies. Their ankles were more highly evolved than those of other reptiles, allowing greater speed and flexibility of movement. Like horses and deer, dinosaurs ran on their toes. In some groups, toes were evolutionarily shed, their number being reduced from the typical five of most reptiles to as few as two functional running toes on some of the faster dinosaur forms.

Sometimes vestigial digits remained, as the dew claw has in some modern mammals. In many cases the leg lengthened and the individual bones of the leg and foot were modified for specific functions. Often when leg and foot bones are found individually they are diagnostic of a certain group of dinosaur. For example, the leg and foot bones of the hypsilophodontid found in Shasta County were clearly those of a fleet-footed animal.

Dinosaurs originally ran on two legs instead of four. These bipedal forms sometimes developed manipulating "hands" for grasping, tearing, and other uses. Over millions of years the outer two of the five fingers became much smaller or nonexistent, and the three remaining fingers evolved so that the animal had a grasping, handlike appendage complete with claws. Tyrannosaurs had just two fingers (the first and second) on their hands, and the strange theropod *Mononykus* had only the first digit left. In the quadrupeds (four-legged dinosaurs) the limbs sometimes became elephantlike, with flattened feet and hooflike toes (see fig. 2.3).

Although the skulls of dinosaurs have many bones and extremely variable configurations reflecting their various modes of making a living, they all have common structures that mark them as truly dinosaurid—and that readily distinguish them from other reptiles and mammals. For example, the large, mammal-like animal called *Dimetrodon,* with a fanlike sail on

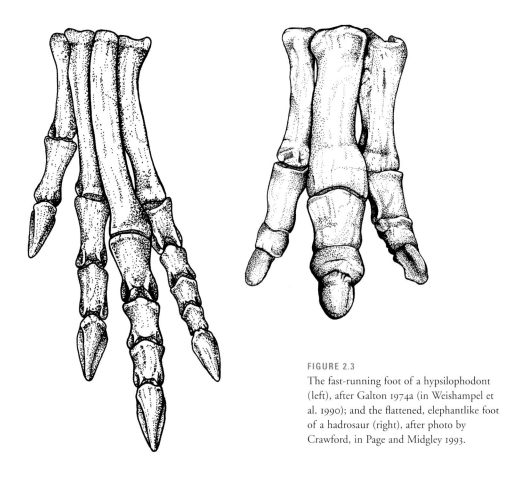

FIGURE 2.3
The fast-running foot of a hypsilophodont
(left), after Galton 1974a (in Weishampel et
al. 1990); and the flattened, elephantlike foot
of a hadrosaur (right), after photo by
Crawford, in Page and Midgley 1993.

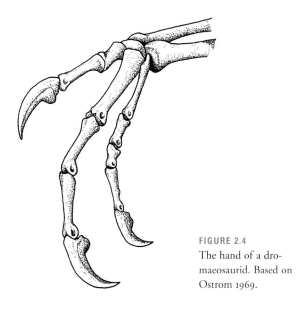

FIGURE 2.4
The hand of a dro-
maeosaurid. Based on
Ostrom 1969.

FIGURE 2.5
The Permian reptile
Dimetrodon is mistak-
enly called a dinosaur.

its back, though often mistakenly called a dinosaur, did not have the structure in the bones of the skull or skeleton that dinosaurs have.

One of the most diagnostic traits separating dinosaurs not only from mammals but also from their reptilian relatives is the numerous and unique positioning of the openings and cavities found in the bones of the skull. In particular, because all dinosaurs are diapsids, they have two pairs of main openings (the temporal fenestrae) behind the eye (see fig. 2.6). Among living reptiles, only the primitive sphenodontid lizards and crocodilians share this trait.

Most dinosaur skulls are a configuration of bones that resemble the open metal structure of some bridges (though there are exceptions, such as *Pachycephalosaurus,* the bone-headed dinosaur). This design, being strong and flexible yet light in weight, has many advantages. Most dinosaurs, moreover, had small brains that did not need to be protected by a solid bony covering (unlike mammals); a small bony braincase nestled within this open framework sufficed. Unfortunately, this delicate skull structure means that complete skulls of most dinosaurs are rarely preserved.

Most of our reconstructions of dinosaurs come from skeletal remains alone, because the nonbony parts of dinosaurs generally decayed before they could fossilize. Only rarely

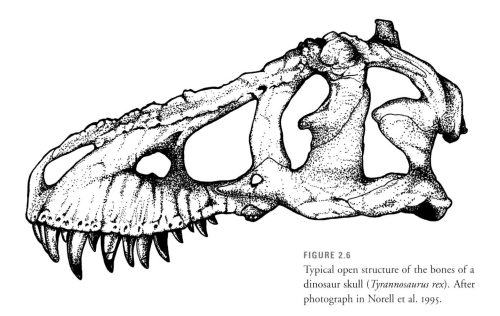

FIGURE 2.6
Typical open structure of the bones of a
dinosaur skull (*Tyrannosaurus rex*). After
photograph in Norell et al. 1995.

do we find impressions of the skin or mummified remains of dinosaurs. Some of the skin
impressions are from hadrosaurs found in Baja California. After death, its desiccated skin
was covered with fine sediment, and upon lithification (conversion to stone) the intricate
details of the scales were preserved. Skin impressions allow artists to recreate the skin texture in dinosaur illustration and sculpture.

DINOSAUR TAXONOMY

The structure of dinosaur hip bones provides the basis for subdividing dinosaurs into two
main groups, or orders: Ornithischia and Saurischia. The hip bones of ornithischians (which
means "bird-hipped") superficially resemble those of birds: the two sets of lower bones, called
the pubis and the ischium, run parallel to each other, extending toward the rear. In most
of the "lizard-hipped," or saurischian, dinosaurs, the pubis and ischium are splayed out, with
the pubis extending forward and the ischium extending backward. These two distinct structures indicate very different ways of attaching muscles in the hip area.

The ornithischians and saurischians share some common features, but they are not that
closely related—they are about as close to each other as are even-toed hoofed mammals (such
as cows and camels) to odd-toed hoofed mammals (such as horses and rhinos). It is inter-

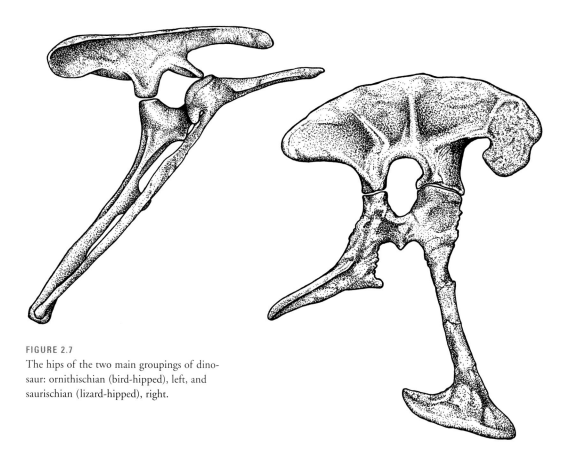

FIGURE 2.7
The hips of the two main groupings of dinosaur: ornithischian (bird-hipped), left, and saurischian (lizard-hipped), right.

esting to note that birds, which many call living dinosaurs, did not evolve from "bird-hipped" dinosaurs but rather from the "lizard-hipped" forms.

Among the saurischians, further evolutionary changes produced the theropods, meat-eating dinosaurs characterized in part by an extra joint in the lower jaw. They had bladelike serrated teeth, and the outer fingers on their hands often were reduced in size or number. Within theropods we see the pubis turn gradually backward with time. Recently discovered fossils indicate that some theropod dinosaurs evolved feathers and structures like feathers, perhaps early on for warmth and protection but (unless secondarily flightless) later for gliding and flight—further evidence of the close link between dinosaurs and birds.

DINOSAUR FINDS IN CALIFORNIA

Figure 2.8 shows the counties where dinosaur fossils have been found in California. In addition, numerous dinosaur fossil remains have been found in the nearby coastal areas of Baja

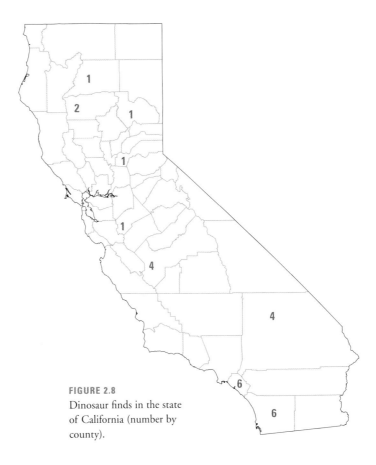

FIGURE 2.8
Dinosaur finds in the state
of California (number by
county).

California, and a single hadrosaur bone was found just north of the California border in
southern Oregon.

The great majority of the dinosaur fossils found in California are from the Cretaceous Period. There are no Triassic dinosaur fossils at all (in part because much of what is now California was ocean at that time), and only three sites have relinquished evidence of Jurassic dinosaurs.

Jurassic Dinosaurs

During the Jurassic, suitable dinosaur habitat existed in much of what is now the eastern
portion of the state. Lush forests likely covered the coastal mountains, while a more arid environment prevailed in the southern interior. Although many types of dinosaurs probably
lived in these environments, Jurassic evidence is meager.

In 1958 James R. Evans discovered dinosaur footprints and trackways in the Early Jurassic Aztec Sandstone of the Mojave Desert in San Bernardino County. Not only are these the

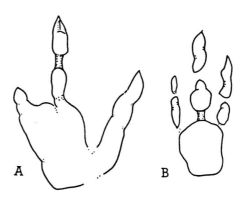

FIGURE 2.9 (ABOVE)
Tracks of the dinosaurs (a) *Grallator* sp. and (b) *Anchisauripus* sp. From Reynolds 1989.

FIGURE 2.10 (LEFT)
Dinosaur tracks in the Jurassic Aztec Sandstone of the Mojave Desert. Photo from the collection of Robert Reynolds.

only known dinosaur tracks from California, but to date they are also the earliest dinosaur evidence from the state. Robert E. Reynolds of the San Bernardino County Museum has analyzed the tracks and says they come from three different small and lightly built theropod dinosaurs: one similar to *Grallator* sp., another to *Anchisauripus* sp., while the third remains unidentified (Reynolds 1989). Both *Grallator* and *Anchisauripus* are ichnogenera, that is, taxonomic categories based only on tracks or trails.

The tracks are impressed in rocks formed from cross-bedded dune sand, indicating that these dinosaurs probably roamed an interior coastal desert environment that was perhaps visited by marine fogs. This moisture allowed sand crescents in the dunes to hold their shape without slumping. As the animals walked through the moist sand, crescent-shaped impact mounds, oriented downslope, formed around each imprint, but as these tracks progressed into dry sand they lost their shape. The fossil prints vary in detail, presumably with the moisture content of the sand and the amount of wind erosion that occurred (Reynolds 1989).

A second site of discovery of Jurassic dinosaur remains is in very late Jurassic marine rocks

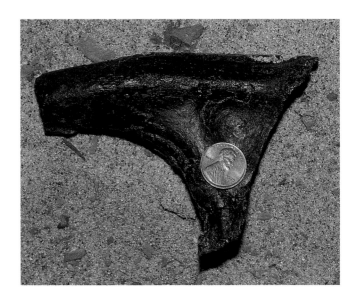

FIGURE 2.11
Jurassic dinosaur rib from the
Sierra Nevada found by Geoff
Christe. Photo by the author.

of the Great Valley Group along the western side of the Sacramento Valley. In late 1998 a rancher, Jim Jensen III, discovered the end of what may be a metatarsal (foot bone) of a medium-sized dinosaur (SC-VR81). John Horner of the Museum of the Rockies (pers. comm. 2001) examined the specimen and noticed that the bone was extremely well vascularized (i.e., it had lots of blood vessels), indicating it was from a small or juvenile dinosaur. The overall shape of the bone suggested that it was not from any known ornithischian dinosaur, but probably from a saurischian, the group that includes carnivorous dinosaurs and sauropods. Because it is so fragmentary, little more can be said about it.

The third area of discovery is in the upper Feather River country of the Sierra Nevada. Here, while doing field mapping in the fall of 1991, geologist Geoff Christe found dinosaur remains in the Late Jurassic Trail Formation (Christe and Hilton 2001). His first discovery, a possible leg bone, is still embedded in the very hard sandstone. In the fall of 1995 Christe found the proximal end (that closest to the spine) of a rib from a dinosaur with a rib cage the size of a bison (see fig. 2.11). He also found several other bone fragments and a tooth fragment in 1997. These are the only Mesozoic terrestrial rocks (formed on land as opposed to under the ocean) to yield reptile bones in Alta California. The bones were found in rocks that Christe interprets to have been sands and gravels of braided stream channels at the time of their deposition. Today these rocks are extremely hard, highly tilted, and slightly altered due to heat and pressure.

Recently, Christie (pers. comm. 2002) has found more bones still encased in their rock matrices. These and the three other tantalizing discoveries give us hope that we will learn more about Jurassic dinosaurs in California.

Compared to the scant record of Jurassic dinosaurs, Cretaceous dinosaur finds in California and adjacent areas are relatively numerous and varied. Dinosaurs from this period have been found from southern Oregon, throughout the length of California, and into northern Baja California. The majority of these were herbivorous dinosaurs.

ANKYLOSAURS Ankylosaurs were armored quadrupeds originating as far back as the Early Jurassic, the skeletal remains of which have been found on every continent except South America (Carpenter 1997a). The bony armor of ankylosaurs—which separates them from all other dinosaurs—consisted of tall spikes, short spines, and keeled plates arranged in transverse bands along the length of the body and tail. The characteristic tail club was made from large terminal plates all fused together. An underlying band of bone was often fused to the armor of the neck. The front legs of ankylosaurs were typically stocky, resembling those of stegosaurs. All four legs were modified to carry the weight of the heavily armored body.

An ankylosaur discovered at Carlsbad, San Diego County, by Bradford Riney of the San Diego Natural History Museum is without doubt one of the most interesting of the dinosaurs found in California. This specimen (SDNHM-33909), found in the 75-million-year-old (Late Cretaceous) marine Point Loma Formation, is one of the most complete dinosaur skeletons ever found in the state. The animal may have drowned in a river or stream before its carcass washed out to sea, where it settled on its back on the shallow ocean floor.

Ankylosaurs are separated into two groups: Ankylosauridae, which had a tail club, and Nodosauridae, which did not. Based on its pelvic armor, the Carlsbad specimen was long considered a nodosaurid (Deméré 1988; Coombs and Deméré 1996), but recently Tracy Ford and James Kirkland (2002) have challenged this interpretation. Although the diagnostic club on the end of the tail was not found, the fossil's armor, tooth morphology, and limb proportions all indicate that the Carlsbad specimen is an ankylosaurid. Ford and Kirkland have named the creature *Aletopelta coombsi*. The genus name *Aletopelta* is derived from the Greek words *aletes*, meaning "to wander," and *pelte*, "shield." The species is named for Walter P. Coombs Jr., who did ground-breaking study of ankylosaurs.

The specimen was relatively intact, giving us a fair picture of what *Aletopelta* looked like when alive. Thomas Deméré (1985) of the San Diego Natural History Museum described the fossil as "lying on its back with the legs splayed out to the sides like some Cretaceous 'road kill.'" Although broken up during its accidental initial excavation, bones were found from every part of its body. Even teeth were recovered. (Teeth are often dislodged from an animal and may be scattered by scavengers or currents.) The specimen retained much of its original armor: patches of bony, interlocking polygonal skin plates about two inches in

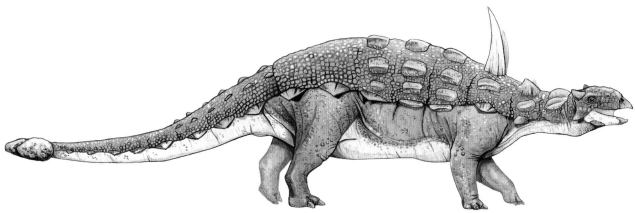

FIGURE 2.12
With a deadly club on the tip of its tail, *Aletopelta coombsi* was covered with heavy armor and spikes. Portions modified from *Euoplocephalus* by Paul 1993; and sketch of *A. coombsi* provided by T. Ford.

diameter, with raised centers. Larger keeled scutes, although not found in place, seem to have been arranged in pairs and most likely ran down the length of the tail.

The fossil ankylosaur found in San Diego County had settled to the bottom of the shallow sea, where the carcass continued to decay. Shark teeth found among the bones suggest that early in this decay cycle sharks fed on the animal. This scavenging probably dismembered much of the carcass and, along with ocean currents, resulted in some loss of the skeleton. Later the remaining bones became a reef for such marine creatures as oysters and rock scallops, a few of which were found still attached to some of the bones. Other invertebrates appear to have burrowed into the spongy cavities of the bones, sometimes giving the bones a deceptively hollow appearance. Numerous depressions carved in some of the bones and dermal armor indicate scavenging by other invertebrates as well (Deméré 1985).

As herbivores, ankylosaurs likely frequented the lush areas around streams and rivers where they would crop the riparian vegetation with their beak and bladelike teeth. This may help to explain why the Carlsbad specimen ended up washed out to sea. Evidence from trackways in Colorado indicates that some species may have been gregarious (Kurtz 2001). Although we will probably never know for sure, it seems plausible that ankylosaurs had protective coloration as do many forest animals of today.

Given their massive bones and durable armor, the preservation of other ankylosaurids is not surprising. In July 1971 in northern Baja California, less than two hundred miles south of San Diego County, Harley J. Garbani found one plate of the armor from an unidentified

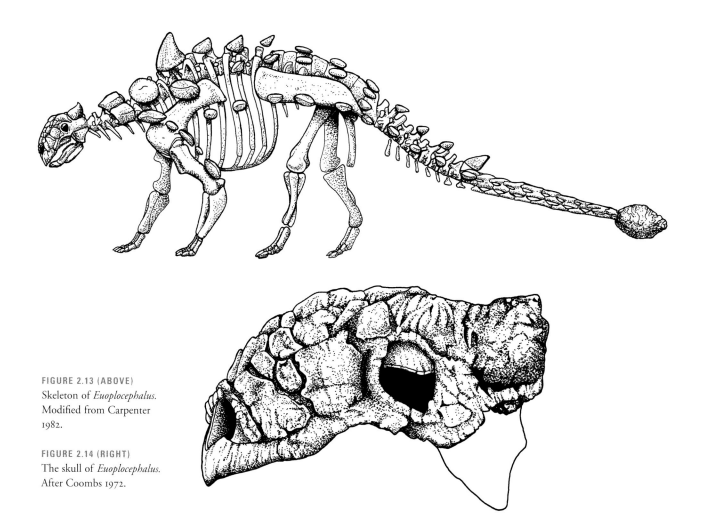

FIGURE 2.13 (ABOVE)
Skeleton of *Euoplocephalus*.
Modified from Carpenter
1982.

FIGURE 2.14 (RIGHT)
The skull of *Euoplocephalus*.
After Coombs 1972.

ankylosaurid (LACM to Mex.-29000). It was in the Late Cretaceous (ca. 73 MYBP) El Gallo
Formation in rocks of comparable age to the Carlsbad find, but whether it was the same type
of ankylosaur we cannot tell.

Hernandez-Rivera (1997) reported another ankylosaurid (probably *Euoplocephalus*) found
in the Late Cretaceous El Gallo Formation as well. Although *Euoplocephalus* means "well-
armored head," virtually the entire body of this creature was wrapped in armor, arranged in
bands interspersed with bony projections of knobs and spikes along its neck and back.

Unlike other ankylosaurs, which tended to be four-toed, *Euoplocephalus* had only three
toes (Carpenter 1982). A slow-moving plant-eater with wide hips and short legs, it must have
plodded along like a living tank. When confronted by a predator or rival, it may well have

FIGURE 2.15

Euoplocephalus had body armor that formed bands interspersed with bony projections of knobs and spikes along its neck and back. Based on skeletal reconstruction by G. Paul, in Lessem and Glut 1993.

used its large, gnarly tail club as a wrecking ball. The club would swing around at high speed as the animal spun its three-ton, eighteen-foot body around (Wallace 1993). Imagine the dull thud and crunch of bone as this club punched into any animal that got in the way.

CERATOPSIANS   *Psittacosaurus,* an early ceratopsian from Mongolia, had neither horns nor a frill like *Triceratops* but is considered a ceratopsian because the bone structure of its face includes the typical beaklike snout and flaring jugals (cheek bones) (Dodson 1997a). According to Dodson (1997b), the Ceratopsia are known only from eastern Asia and North America. They originated in Asia, later migrating to North America where divergence progressed.

The best known of the ceratopsians (meaning "horn faced") is *Triceratops,* named for its three horns. It also had a parrotlike beak and a massive frill. The horns and frill were likely used for defense and sexual display. Large-horned ceratopsians like *Triceratops* compose the Neoceratopsia, a diverse quadrupedal group of creatures that were up to twenty-eight feet in length. Ceratopsians like the so-called duck-billed dinosaurs developed complex dental batteries in which the teeth were locked together in vertical columns and longitudinal rows and were constantly replaced as they wore along the cutting edge (Ostrom 1966; Forster and Sereno 1997). These chopping teeth, along with the beak, suggest a vegetarian diet. These animals also had large nasal openings in their enormous skulls. The largest species of ceratopsians had the biggest skulls of any land animal to have ever lived, some over eight feet long. The large ceratopsians were some of the last dinosaurs to appear in the fossil record.

The only evidence of this group in California comes from the remains of an unidentified ceratopsian found in the Late Cretaceous El Gallo Formation of Baja California (Hernandez-Rivera 1997). Until more complete fossil specimens are found we can't be sure which cera-

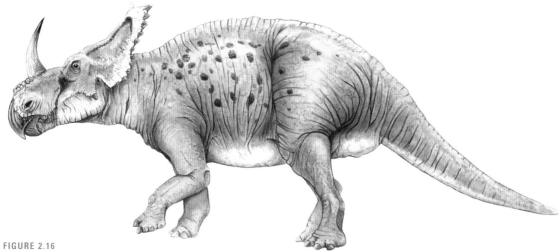

FIGURE 2.16

Monoclonius was a typical ceratopsian, but unlike the more familiar *Triceratops* it had only one horn. Based on the reconstruction of *Centrosaurus (Eucentrosaurus)* by Gregory S. Paul, in Dodson 1996.

topsian this was, or what other ceratopsians may have lived in Baja or Alta California. The relative lack of ceratopsians in California suggests that the environment here in the Late Cretaceous was different from the localities of the upper Midwest and Canada, where *Triceratops* is perhaps the most abundant Late Cretaceous dinosaur fossil.

In Canada, the ceratopsian *Centrosaurus* has been preserved en masse, which suggests it may have been a herding animal (Currie and Dodson 1984). Single-species bone beds indicate that animals died together crossing a river or in a catastrophic flood, or were buried as a herd from falling volcanic ash. *Triceratops* has not been found en masse, suggesting that it may simply have been an abundant but rather solitary animal (Dodson 1997d).

HADROSAURS   Hadrosaurs, the so-called "duck-billed" dinosaurs, were common in the Late Cretaceous. They were large, usually bipedal, herbivorous animals that may have occasionally walked on all fours (perhaps when foraging). They had a toothless, ducklike snout, but toward the rear of the jaws was a battery of teeth that formed grinding surfaces capable of reducing coarse vegetation to digestible mash (see fig. 2.17). Marvelous skin impressions from the Baja hadrosaur *Lambeosaurus laticaudus* (UCMP-137303) tell us the sizes, arrangement, height, and even texture of this animal's scales and give a beautiful picture of what it looked like when it was alive.

Although we know more about aspects of hadrosaurid life history than about the life histories of any other group of dinosaurs (Forster 1997), the behavior of the specific hadrosaurs

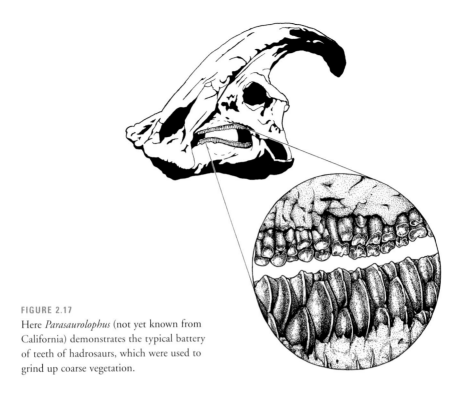

FIGURE 2.17
Here *Parasaurolophus* (not yet known from California) demonstrates the typical battery of teeth of hadrosaurs, which were used to grind up coarse vegetation.

that roamed the west coast remains unknown. Horner (1997a) points out that some hadrosaurs were probably herding animals. They laid eggs in communal nest sites like some present-day ratites (ostriches and emus) and frequented the same nest sites year after year. Some may have also protected their young, much as crocodiles, many birds, and mammals do today. There is also indirect evidence that the Baja California hadrosaurs were preyed on or scavenged upon by carnivorous dinosaurs, for the teeth of theropods were found among the scattered bones of at least two individuals in Baja (Morris 1981).

Hadrosaurs were perhaps the most common dinosaur to have inhabited the Pacific coastal region of California, and indeed, more fossil hadrosaurian remains have been found in California than those of any other dinosaur. More generally, fossils of hadrosaurs have been found all along the west coast: in Alaska, southern Oregon, throughout the length of Alta California, and on into northern Baja California (Rich et al. 1997). They are also known from other parts of North America, Central and South America, and Eurasia (Forster 1997).

The relative abundance of hadrosaurian fossils on the west coast could be attributable more to the habits of these animals than to large population sizes. Morris (1981) suggests that the Baja species *Lambeosaurus laticaudus,* and perhaps even *Saurolophus* from Alta Califor-

FIGURE 2.18
Hadrosaur remains found in
California (number by county).

nia, may have frequented the shallower areas of bays, lagoons, and estuaries, likely places for burial and fossilization. He likewise contends (1973a) that the hadrosaur's long, laterally flattened tail may have been used for swimming in such aquatic environs. Ostrom (1964), however, points out that criss-crossed tendons in the tail more likely functioned not for swimming but, as in many other dinosaurs, for balance while walking and running. Although the flattened tail could have been used for swimming across rivers and swamps, it did not necessarily indicate a primarily aquatic lifestyle.

Other evidence suggests that hadrosaurs lived in lowland forested areas, as well as in forested highlands. Although hadrosaur remains in Baja California have been found mainly in coastal fluvial deposits, whether they lived there or were washed down rivers to that environment is open to speculation.

Unfortunately most of the hadrosaurian remains found in California are so fragmentary that all that can be said of them is that they were hadrosaurs. The only exceptions are two

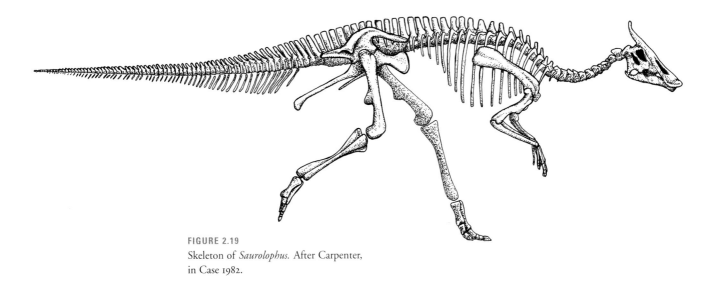

FIGURE 2.19
Skeleton of *Saurolophus.* After Carpenter,
in Case 1982.

nearly complete skeletons of the hadrosaur *Saurolophus* from the west side of the San Joaquin Valley (the most complete dinosaur fossils ever found in California), and enough bones in Baja to describe the huge hadrosaur named *Lambeosaurus laticaudus.*

The name *Saurolophus* means "crested reptile." It is distinguished from other hadrosaurs by a small, bony, spikelike crest that projects up and back from the top of the skull (see fig. 2.21). This spike—an extension of the nasal bones, connected to the nasal area by hollow passages—may have functioned in sexual display and perhaps defense (Dodson 1975), but it also may have been used to amplify and resonate sound (Weishampel 1981a). An additional distinguishing feature of the skull is a cavity hollowed out of the facial bones in front of the eyes. This cavity, too, is connected to the nasal air passages and may have been covered by a fleshy pouch that was inflated for resonating sound (DeCourten 1997).

Reaching lengths of about thirty to forty feet, *Saurolophus* was a bipedal, slender-legged plant-eater that may have gone down onto all four of its hoofed feet while feeding close to the ground. It had a toothless beak used for cropping plants and a battery of grinding teeth in its jaws for the processing of coarse vegetation.

The two California specimens of *Saurolophus*—both from the Late Cretaceous Moreno Formation and about 70 million years old—were excavated by Chester Stock's crews in the Panoche Hills of Fresno County in 1939 (LACM/CIT-2760) and 1940 (LACM/CIT-2852). (*Saurolophus* has also been found in Montana, Alberta, and even Mongolia.) The first California specimen comprised most of the skull plus mandible, pelvis, and limb bones; however, the bones were in a poor state of preservation. The second was better preserved, con-

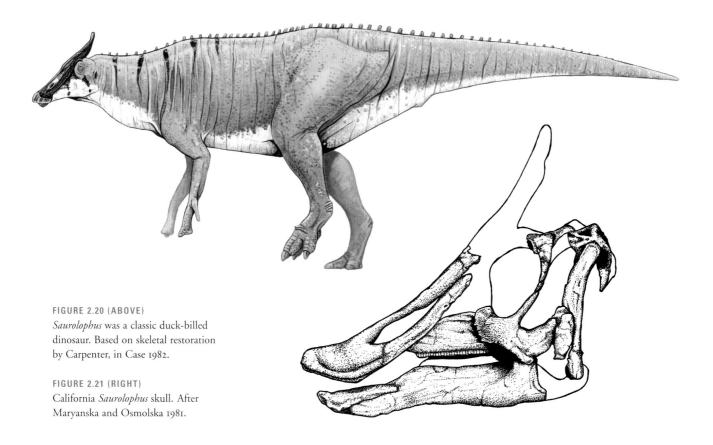

FIGURE 2.20 (ABOVE)
Saurolophus was a classic duck-billed dinosaur. Based on skeletal restoration by Carpenter, in Case 1982.

FIGURE 2.21 (RIGHT)
California *Saurolophus* skull. After Maryanska and Osmolska 1981.

sisting of a nearly complete skeleton with all of the mandible and most of the skull. Both skulls lacked the most important diagnostic part, the postnarial crest (at the top of the skull, above the nostrils), but they did have the elongate, spatulate premaxillae ("duck" bills) complete with narial passages (Morris 1973a).

Numerous remains of hadrosaurs have been found in the extension of the Peninsular Range that runs south from San Diego into Baja California. These remains are all from the Late Cretaceous El Gallo and La Bocana Roja Formations. The first of these hadrosaur discoveries was in 1953 when J. Wyatt Durham and Joseph Peck of the University of California, Berkeley, found foot bones from two small individuals (UCMP-43251) in the El Gallo Formation. Langston and Oakes (1954) described these bones as from a hadrosaur about the size of *Kritosaurus,* a hadrosaur about twenty-five to thirty feet in length. Many remains of hadrosaurs have subsequently been found in this area, but none from *Kritosaurus,* so perhaps they are from a lambeosaur or even *Saurolophus.*

A hadrosaur described by Morris (1981) as *Lambeosaurus laticaudus* (LACM-17715) was discovered by Morris's crews in Baja California in the summer of 1966. This species has a

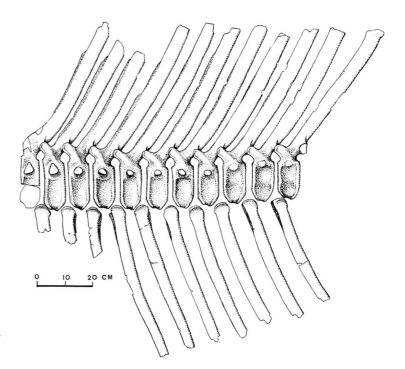

FIGURE 2.22
Laterally compressed spine
of *Lambeosaurus laticaudus*.
From Morris 1981.

specialized tail that is laterally compressed with long neural spines and haemal arches projecting from the vertebrae (see fig. 2.22). The specimen was described originally as *Hypacrosaurus altispinus* (Morris 1967c), in part because of the bone structure of the tail. The later discovery of a partial skull suggested it was a lambeosaur, for whereas *Hypacrosaurus* has a closed narial canal on the premaxilla with only the distal part open, all species of *Lambeosaurus* have an open narial canal (Morris 1972).

One of the most odd and distinguishing features of lambeosaurs is a hollow, vertical crest on the head—hence their designation as "hollow crested" hadrosaurids (see fig. 2.23). Nasal passages run through this crest, again suggesting that it may have functioned not only in visual display but also as a resonance chamber for vocalization (Weishampel 1981a). Like *Saurolophus* in California, the Baja lambeosaur skull was missing the crest. Behind this crest in most lambeosaurs is a smaller, narrow, and sharp-pointed crest that may have joined a ridge of skin that ran along the animal's back.

Lambeosaurus laticaudus was a huge hadrosaur, with many of the Baja remains indicating an average length of thirty feet—though some of the bones discovered in Baja (LACM-26757) suggest an animal more than fifty feet in length (Morris 1972), which would make it one of the largest hadrosaurs ever found. According to DeCourten (1997), this specimen might have weighed more than twenty tons. The fossils of *Lambeosaurus laticaudus* come

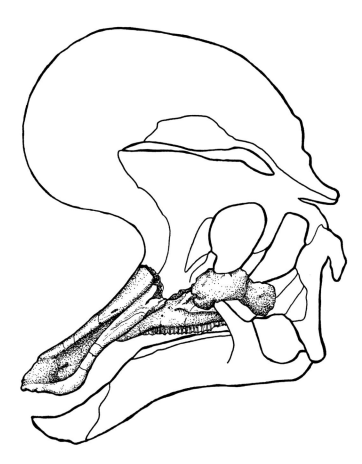

FIGURE 2.23
Lambeosaurus laticaudus
skull, showing portions dis-
covered in Baja California.
Outline of skull after
Weishampel 1981b.

from river and floodplain deposits that are a little older than 70 million years (Morris 1981). Other lambeosaurs have been found in Montana and the provinces of Alberta and Saskatchewan in Canada.

Besides the fossils found in Fresno County and Baja, other fragmentary California hadrosaur remains have been discovered in San Diego (SDNHM-35342, 67368, 66640), Stanislaus (UCMP-32944), Orange (OCNHF-1785), and Tehama Counties (YPM-PU-19333). No scientific paper was ever written on the Tehama County discovery, although Horner (1979) does mention its existence in his paper on dinosaur remains found in marine environments. It is interesting to note that, because they are so fragmentary, none of the California specimens, even those in the southern part of the state, have been referred to as lambeosaurs. But surely, with the Baja deposits less than two hundred miles from San Diego, lambeosaurs were in Alta California as well.

How *Saurolophus* and *Lambeosaurus laticaudus* survived predation in California is difficult to surmise, since they have no obvious means of defense. Their unique vocalization may have

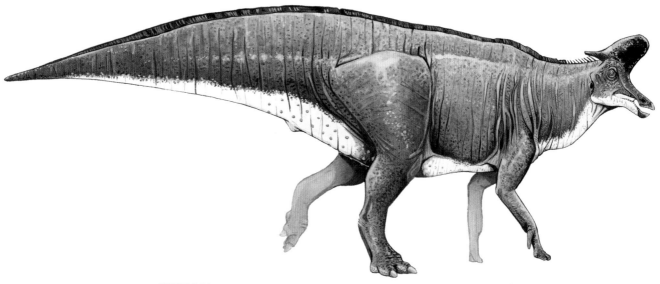

FIGURE 2.24
Lambeosaurus laticaudus is one of the largest of the duck-billed dinosaurs ever found. Modified from skeletal illustration of *Corythosaurus*
by G. Paul, in Paul 1987.

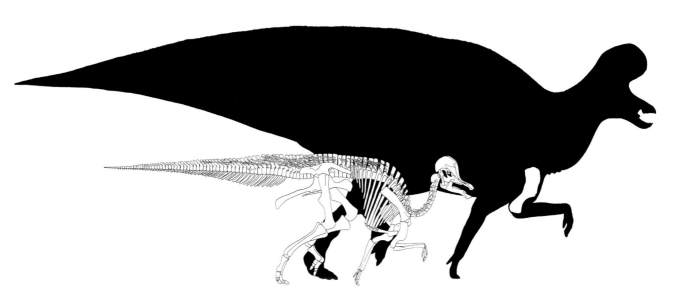

FIGURE 2.25
Lambeosaurus laticaudus in black silhouette with Baja humerus, and
skeleton of average-sized hadrosaur *Corythosaurus. Corythosaurus*
modified from Paul 1987.

served as a warning for individuals in relatively close proximity. If they were true herding animals, collective alertness and perhaps collective defense postures and actions may have come into play. Whether they lived in forested or open areas, they may have had camouflaging coloration. If Morris is correct, both *Saurolophus* and *Lambeosaurus laticaudus* may have spent much of their time in aquatic environments such as coastal lagoons and estuaries, perhaps avoiding predation in much the same way hippos do today.

Then again, the fact that hadrosaurs laid many eggs and the young grew quickly suggests that it may not have been their defenses but rather their sheer numbers that ensured the survival of the species (Horner 1997a).

SMALL HERBIVORES   Several types of relatively small herbivorous dinosaurs could have inhabited California during the Cretaceous, but so far we have evidence of only one. In 1991 in Shasta County the author and his son, Jakob, along with geologist Tom Peltier discovered foot and lower leg bones of a small hypsilophodontid dinosaur (SC-VRO4; Hilton et al. 1997). The bones were retrieved from two concretions in 115-million-year-old, Early Cretaceous marine rocks of the Budden Canyon Formation, near Redding (Murphy et al. 1969). Generally speaking, these fossils are the oldest identified Cretaceous dinosaur remains on the Pacific coast of the United States and Canada outside Alaska (Rich et al. 1997). Hypsilophodonts have been found in rocks ranging from Middle Jurassic through to the end of the Cretaceous (170–65 MYBP). Their fossils have been found in Eurasia, Africa, Australia, Antarctica, and North and South America (Sues and Norman 1990).

Comparison of the bones from Shasta County to the same bones of the hypsilophodontid *Thescelosaurus,* a relatively small plant-eating dinosaur, showed them to be very similar. Hypsilophodonts had slender hind limbs and smaller front limbs, which may have been used for grasping during browsing. Unlike the lumbering dinosaurs we often think of, hypsilophodonts were fleet-footed runners (Brett-Surman 1997); their foot and leg bones are reminiscent of today's fast-running plains animals, with long, closely spaced metatarsals (middle foot bones) and a tibia (lower leg) longer than the femur (upper leg). Hypsilophodonts had a long tail, and like the tails of many other ornithischians it was partially stiffened by ossified tendons and probably acted as a stabilizer and counterbalance when running (Sues 1997). Hypsilophodonts had chisel-shaped cheek teeth and retained the premaxillary teeth (those forward of the cheek teeth) (Brett-Surman 1997). One of the hypsilophodonts, *Hypsilophodon,* was probably a herd animal, as several have been found together in one fossil site (Czerkas and Czerkas 1991). Some hypsilophodonts may even have used colonial nesting sites, but the young probably had to fend for themselves after hatching (Sues 1997).

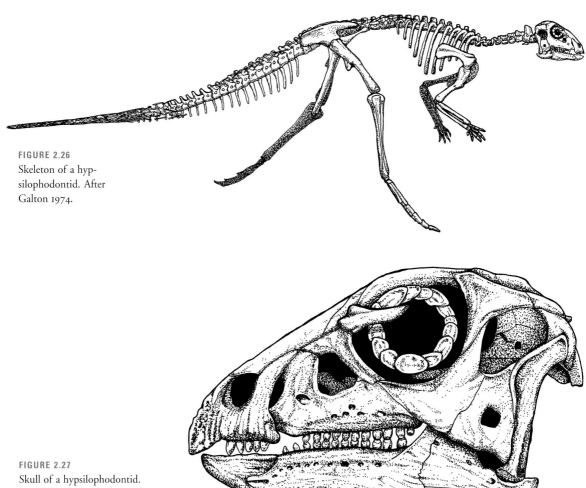

FIGURE 2.26
Skeleton of a hyp-
silophodontid. After
Galton 1974.

FIGURE 2.27
Skull of a hypsilophodontid.
After Galton 1974.

FIGURE 2.28
Hypsilophodontid dinosaur,
a small, fleet-footed herbivore. Based on skeletal
restoration by Galton 1974.

The hypsilophodont from Shasta County probably lived in a fairly rugged coastal forest, so it may have been somewhat solitary, using its agility and quick speed to avoid predation, much like deer of today. It may also have had a camouflaged skin pattern to help it blend with the vegetation (somewhat similar to green iguanas in the Central American jungles). Redwood foliage, ferns, leafy trees, and cycads are some of the more common fossils found in layers adjacent to the hypsilophodont, hinting at the food sources available to this herbivore and at the makeup of the forest it inhabited.

Cretaceous Carnivores

To date, very little evidence of carnivorous dinosaurs has been found in Alta California, but Baja California has surrendered more abundant remains, suggesting the types of carnivores that may have roamed California and preyed upon its hadrosaurs, ankylosaurs, and other herbivores.

The first evidence of a theropod (meat-eating) dinosaur in California was discovered in 1994 east of Sacramento in the community of Granite Bay, when Patrick Antuzzi found the midsection of a long bone while searching through rocks from a subdivision ditch excavation. Gregory Erickson (pers. comm. 2000), a paleobiologist from Florida State University, examined thin sections of the bone (VRD57C and VR57D) under a polarizing light microscope and concluded, based on the microstructure of the bone, the pattern of vascularization (blood supply), and the bone's overall structure, that the bone was almost certainly from a young theropod dinosaur.

Fossils of plants, seeds, and land snails found in the same deposits as this dinosaur, as well as other geologic evidence gleaned from the sediment, paint a vivid sketch of that Late Cretaceous Sierra Nevada forest (Hilton and Antuzzi 1997). The landscape was like the rugged Oregon coast, with an actively eroding coastline in front of a lush forest. The forest would have been very different from today's, however, with many primitive, leafy, flowering trees (some much like present-day magnolias) and coniferous trees that today are found as natives only in the southern hemisphere (some resembling Norfolk Island pine and monkey puzzle trees). The understory would have included tree ferns (see fig. 2.29), seed ferns, horsetails, and cycads. Land snails were also found in these deposits (Roth 2000). Dinosaurs would have roamed the forest like deer and jungle elephants, and the theropods, like forest tigers, were their predators.

About two hundred miles to the south of San Diego in northern Baja California are remains of other Late Cretaceous theropod dinosaurs. These theropods might be suggestive of the types of animals that may have roamed the Peninsular Range of southern (Alta) California, as well as perhaps even in the northern part of the state. The following carnivorous dinosaurs were found in Baja California (Hernandez-Rivera 1997): *Albertosaurus, Labocania*

FIGURE 2.29
Trunk of a tree fern found
by Patrick Antuzzi and
the author in the Late
Cretaceous Chico Formation
at Granite Bay. Photo
by the author.

anomala, Troödon formosus, and *Saurornitholestes.* In addition, two indeterminate forms, a tyrannosaurid and a dromaeosaurid, may or may not be the same as dinosaurs already found, and further material suggests that an ornithomimid was also present.

CARNIVOROUS DINOSAURS   Among the meat-eating varieties of dinosaur were large, big-headed theropods with small arms and strong legs. A good example is the genus *Albertosaurus,* a more lightly built, smaller relative of *Tyrannosaurus,* named for the province of Alberta, Canada, where it was first found. Its fossils have also been found in Late Cretaceous deposits in Baja California (Hernandez-Rivera 1997) and Montana (Dodson 1997b). An as-yet-undescribed example may have been found in Alabama (Dodson 1997b), and Clemens and Nelms (1993) reported a small *Albertosaurus* from the North Slope of Alaska. Having been found as far north as Alaska and south as far as Baja, it seems safe to presume that *Albertosaurus* lived in Alta California as well.

Albertosaurus was a bipedal carnivore with muscular hind legs and small, comparatively weak forelimbs, each sporting two-fingered "hands" complete with claws. Small sets of horn-like projections of bone were on top of the skull: a larger set in front of the eye and smaller ones in back (Carpenter 1997b; see fig. 2.31). Adults were about twenty-five feet long and

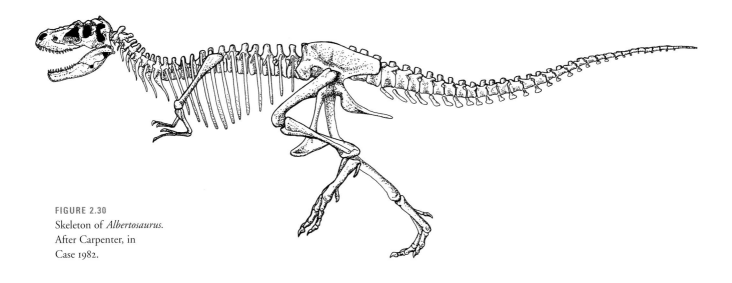

FIGURE 2.30
Skeleton of *Albertosaurus.*
After Carpenter, in
Case 1982.

had strong jaws lined with sharp, serrated teeth mounted in a wide-muzzled skull that was unlike that of any other tyrannosaurid (Lessem and Glut 1993). The skull and teeth were perfectly built for ripping the flesh from its victims. Its eyes were forward-looking enough to have given it fair binocular vision and thus substantial depth perception, critical in hunting (McGowan 1983a).

Albertosaurus is thought to have hunted *Maiasaura* and other duck-billed dinosaurs (Wallace 1993). Like monster wolves, they may have hunted in packs in their northern haunts (Currie 1997b). In the California region hadrosaurs were likely on their list of prey, because in Baja, teeth of carnosaurs were found among the scattered bones of at least two hadrosaurs (Morris 1981).

Other unidentified carnosaur remains have also been found in northern Baja, mostly in the form of the occasional tooth (Hernandez-Rivera 1997). Many of these were probably from *Albertosaurus,* but there may be other types as well. Rodríguez-de la Rosa and Aranda-Manteca (1999) report a "scavenging" theropod from an unknown family in the El Gallo Formation of Baja. It is represented by a tooth that is unlike any other theropod tooth yet found.

Labocania anomala, an animal about two-thirds the size of *Tyrannosaurus rex* (Molnar 1974), was found by Harley J. Garbani in northern Baja California in the summer of 1970. It was described by Ralph Molnar (1974) from fragmentary skull and postcranial material (LACM to Mex.-20877/JHG [HJG] 65) as a new genus of theropod. *Labocania anomala* gets its name from the Late Cretaceous La Bocana Roja Formation of northern Baja. This formation is Late Cretaceous but older than the overlying El Gallo Formation, which is about

FIGURE 2.31 (RIGHT)
Skull of *Albertosaurus.* After
Russell 1970.

FIGURE 2.32 (BELOW)
Albertosaurus, a lightly built
carnosaur that was a smaller
relative of *Tyrannosaurus rex.*
Based on the skeletal restoration by Paul 1988.

73 million years old (Kilmer 1963). The cranial elements of *Labocania anomala* are relatively more massive than in other theropods (Molnar 1974), and the skull, with its thick
snout and jaw, is more like those of Asian carnosaurs than ones found in North America
(Lessem and Glut 1993).

FIGURE 2.33
Labocania anomala was a fierce carnosaur about two-thirds the size of *Tyrannosaurus rex.*

TROÖDONTIDS Northern Baja California has also yielded evidence of the theropod *Troödon formosus,* in the form of a tooth. *Troödon formosus* was described by Joseph Leidy in 1856, making it the first dinosaur in the western hemisphere named and still called by its original name (Varricchio 1997). *Troödon formosus* has been found in Alberta, Mongolia, and, in the United States, Montana and New Mexico (Norell et al. 1995; Varricchio 1997). Nessov and Golovneva (1990) report *T.* cf. *formosus* from the Russian Arctic. This wide range suggests that, like *Albertosaurus, Troödon* may also have lived in Alta California.

Only six to eight feet long, *Troödon* was an agile hunter. Agility is evident in its lightly built body and long, delicate hind limbs complete with large toes, indicating that it was capable of long strides while running (Osmolska and Barsbold 1990). Although each hind foot had three toes, one claw rotated upward and was held off the ground, so the foot essentially ran on just two toes, an efficient trait promoting speed (Varricchio 1997). *Troödon*'s strong hands were probably capable of fairly precise dexterity (see fig. 2.34). It could rotate its front limbs, each of which was equipped with three sharp claws, the second digit being especially formidable (Osmolska and Barsbold 1990). It is evident from the body structure that *Troödon* was an efficient hunter of relatively small prey such as lizards, dinosaur hatchlings, mammals, and perhaps even insects (Osmolska and Barsbold 1990).

Several features of the *Troödon* skeleton suggest that Troödontidae are closely related to ornithomimids as well as dromaeosaurids (Varricchio 2001). Some evidence also suggests that the Troödontidae may be more closely related to *Archaeopteryx* than are modern birds, and that like modern ratites (ostriches, emus, etc.) they may have evolved from flighted ancestors (Longrich 2001).

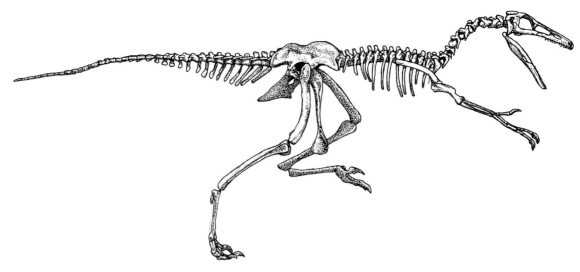

FIGURE 2.34
Skeleton of a troödontid. Postcranial skeleton modified from photo by
Selyem, in Lambert 1993. Skull: orbit and posterior cranial material
modified from *Stenonychosaurus inequalis* from Currie 1985; snout
modified from *Saurornithoides junior,* as illustrated by Sabbath. Both
skull references from Osmolska and Barsbold 1990.

Troödon had the largest brain size (relative to body size) of all known dinosaurs (Barsbold 1997). It had a long, narrow skull and a delicate jaw with long and closely spaced, sharp serrated teeth (see fig. 2.35). Hence the name *Troödon,* meaning "wounding tooth" (Wallace 1993). Tooth counts in *Troödon formosus* run as high as 122 in one individual, perfect for cutting flesh or grasping small prey (Varricchio 1997). Its relatively large eyes were directed forward, providing good depth perception, and the large eye size may mean that it was a nocturnal hunter. The presence of a periotic sinus and an enlarged middle ear cavity indicates that it had a keen sense of hearing, possibly with the ability to detect sounds of very low frequency (Osmolska and Barsbold 1990). In contrast, small external nostrils and comparatively narrow, yet long, olfactory tracts indicate that its sense of smell was not advanced (Osmolska and Barsbold 1990).

Troödon laid clutches of up to twenty-four eggs in a low, earthen mound (Varricchio 1997). The eggs appear to have been laid two at a time several days apart, and in an upright (vertical) position. The mounded earthen nest may suggest that one or both parents took care of the eggs and young in a manner similar to ostriches, with the brooding adult likely maintaining a body temperature that sometimes exceeded the ambient air temperature (Varricchio 1997, 2000).

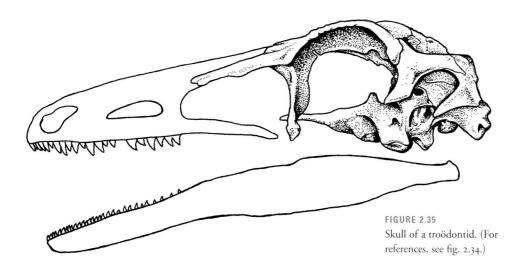

FIGURE 2.35
Skull of a troödontid. (For
references, see fig. 2.34.)

FIGURE 2.36
Troödon formosus with insu-
lating feathers. (For references, see fig. 2.34.)

DROMAEOSAURIDS   The distinguishing characteristic of dromaeosaurids is the large, knifelike
raptorial claw on the second toe of the hind foot, which was normally held off the ground.
These dinosaurs also had a more birdlike pubis than any other known dinosaur. Their front
limbs were long, with flexible wrists and "hands," each having three lengthy fingers with long,
curved claws. Though small animals, they must have been formidable hunters.

FIGURE 2.37
Saurornitholestes had a skele-
ton similar to that of
Velociraptor and may have
been a feathered dinosaur.
After Ostrom 1969.

At least one dromaeosaurid, *Saurornitholestes,* may have lived in Alta California, but as yet the only evidence comes from Baja California, in the form of two teeth found in the Late Cretaceous El Gallo Formation (LACM to Mex.-42637/HJG 689 and 42675/HJG 696). It is also known from Late Cretaceous rocks of Alberta, Canada (Currie 1997c), and from western Montana (Horner 1997b).

Saurornitholestes, which means "lizard bird-thief" in Greek, is currently thought to be a velociraptorine dromaeosaurid. Only about six feet long, it was a lightly built carnivore with a relatively large head and narrow snout. Its skeleton is similar to that of *Velociraptor mongoliensis,* which was featured larger than life in the movie and in Michael Crichton's novel *Jurassic Park* as an intelligent, vicious, perhaps warm-blooded animal that hunted in packs.

Today there is great interest in the dromaeosaurids because of their close relationship to birds (Currie 1997a). As has been suggested for the Troödontidae, these animals, too, may have evolved from flighted ancestors (Longrich 2001). A recent discovery in Lianoning Province, China, of an eagle-sized dromaeosaur called *Sinornithosaurus millenii* is complete with integumentary structures (downy featherlike structures) and a birdlike wishbone, further evidence of the link between dromaeosaurids and birds. Although this dinosaur did not fly (or was secondarily flightless), it had evolved the prerequisites for powered flight in its shoulder girdle, and is the most birdlike of all of the dinosaurs discovered thus far (Xu, Tang, and Wang 1999). Recently Norell et al. (2002) report a nonavian dromaeosaurid dinosaur from the Early Cretaceous Jiufotang Formation in China to have pinnate feathers, complete with rachises and barbs, on both its front and hind limbs as well as feathers extending over seven inches from the last

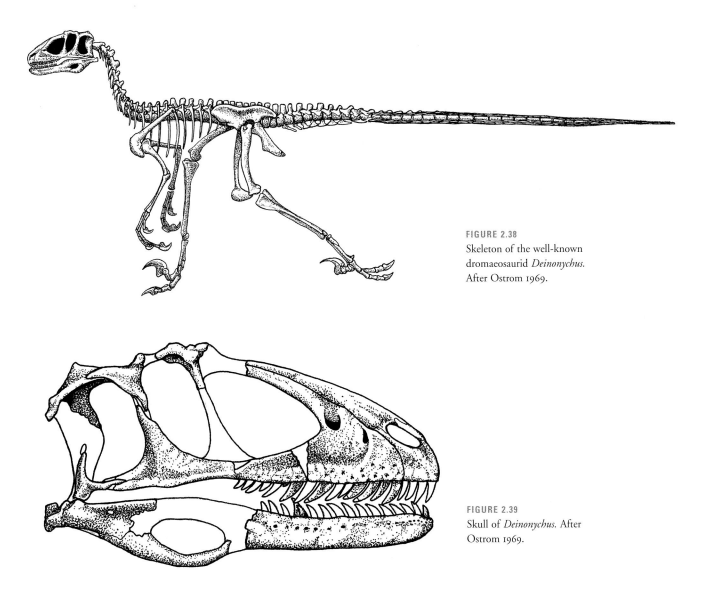

FIGURE 2.38
Skeleton of the well-known
dromaeosaurid *Deinonychus.*
After Ostrom 1969.

FIGURE 2.39
Skull of *Deinonychus.* After
Ostrom 1969.

vertebra of the tail. The feathers are identical to those found on modern birds, leading Norell et al. to conclude that feathers evolved on dinosaurs before the emergence of birds.

Other unidentified remains of dromaeosaurids have also been found in northern Baja, which could be *Saurornitholestes* or other types. Perhaps future discoveries will add to the list of dromaeosaurids found in the California region.

ORNITHOMIMIDS   Ornithomimids had comparatively big eyes and a large brain in a small, birdlike skull atop a long, serpentine neck. The name ornithomimid means "bird mimic,"

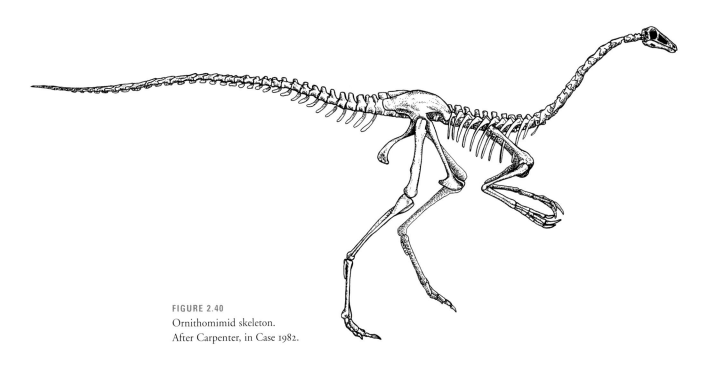

FIGURE 2.40
Ornithomimid skeleton.
After Carpenter, in Case 1982.

and the group is generally known as the "ostrich dinosaurs." Their leg bones—a tibia and fibula longer than the femur of the upper leg—plus other structural changes suggest that they were the fastest of all dinosaurs (Currie 1997c). Their slender, muscular legs provided the swiftness and agility to hunt lizards, insects, and perhaps even small mammals. A typical ornithomimid, such as *Gallimimus,* could be thirteen feet in length with long, thin "arms," each with three thin, clawed fingers, perfect for manipulating prey and perhaps for digging them out of their hiding places (see fig. 2.40).

One unidentified species of ornithomimid was found in northern Baja California, about two hundred miles south of San Diego; it is from the Late Cretaceous (Hernandez-Rivera 1997), but ornithomimid fossils are known from rocks as old as the Late Jurassic (Barsbold and Osmolska 1990). Like the Baja specimen, northern ornithomimids, which have been found in Europe, Asia, and North America (Osmolska 1997), are usually found in lowland sediments associated with lush and warm temperate or possibly even subtropical climates.

Although ornithomimids have beaks instead of the sharp teeth of a predator (see fig. 2.41), their remarkable physical similarity to other known meat-eaters and the lack of a mechanism to grind vegetation makes most scientists think they were predatory (Osmolska et al. 1972; Russell 1972; Barsbold and Osmolska 1990). However, as in tortoises and ostriches, a beak can also successfully crop vegetation, so they could have been herbivores (Nicholls and Russell 1985).

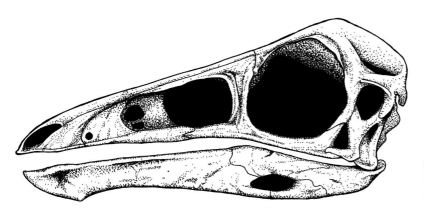

FIGURE 2.41
Ornithomimid skull. After
Carpenter, in Case 1982.

FIGURE 2.42
Shown here as partially
feathered, ornithomimids are
known as the "ostrich dinosaurs." Modified from the
skeletal restoration of *Orni-
thomimus velox* by Paul 1988.

Or perhaps they were omnivorous, eating both animal and vegetable matter (Osmolska 1997). Recently Norell et al. (2001) reported two new ornithomimid dinosaurs discovered in Canada and Mongolia in which the soft tissue of the beaks are preserved, apparently showing structures used for straining sediment. Today certain ducks and flamingos have similar structures. Whether these ornithomimids are the norm or specialized forms we do not yet know.

If ornithomimids were not themselves predators, their agility would have been not for hunting but for escaping swift predators.

OTHER THEROPODS   Fossil evidence from northern Baja California suggests that other theropods may have been present in Baja and, by extension, Alta California as well. Interestingly, evidence from a single tooth crown (FCM06/053) found in the Late Cretaceous El Gallo Formation of Baja California suggests that at least one of these theropods may have used venom to help kill its prey (Rodríguez-de la Rosa and Aranda-Manteca 2000). The blade-like tooth is labio-lingually compressed and the posterior denticles are located within a longitudinal groove that runs over the two distal thirds of the posterior carina. Similar structures, which may have evolved to conduct an oral toxin, are found in elapid snakes (such as cobras and coral snakes), helodermid lizards (such as Gila monsters), other extinct Mesozoic vertebrates, and some Late Cretaceous varanoid lizards (relatives of the Komodo dragon). If in fact this theropod was venomous, this would be a case of the fossil record mimicking the fantasy of moviemaking: Spielberg's *Jurassic Park* featured a venomous dilophosaur.

OTHER CRETACEOUS TERRESTRIAL REPTILES

It was not just dinosaur remains that floated down creeks and rivers to be deposited off the coast of Mesozoic California. Other terrestrial reptiles encountered the same fate, including a land-dwelling tortoise found in the Great Valley Group of Alta California. However, it is Baja California, with its freshwater and lagoonal deposits, that offers most of the evidence we have for non-dinosaur terrestrial reptiles. Some of these rocks were originally deposited in braided streams that entered the sea, whereas the lagoonal strata were probably laid down where spits and bay mouth bars separated protected water areas from the force of ocean waves.

Most of the Baja California terrestrial remains are from crocodilians. Although crocodilians spent most of their time in the water, perhaps even saltwater lagoons, they are considered terrestrial, as opposed to marine, because they did not live in the ocean proper. While we may find it hard to imagine the activities of a dinosaur, the crocodiles and alligators of today give us an idea of how their ancestors lived.

Several crocodilians have been found in the Late Cretaceous El Gallo Formation of northern Baja California. Among these specimens are *Brachychampsa* and *Leidyosuchus,* which are well known from Cretaceous rocks of the western interior of North America. *Brachychampsa* was a specialized alligator that crushed and ate animals with hard shells (Carpenter and Lindsey 1980). The jaws of *Brachychampsa* are more robust than in other crocodilians, and its teeth, which are short with wrinkled crowns, are stronger and blunter than the typical sharp teeth of most alligators (Carpenter and Lindsey 1980). It was a tooth (LACM-101159/28992) of *Brachychampsa* that was found in Baja.

Like alligators and unlike crocodiles, which have V-shaped snouts, *Brachychampsa* had broad jaws shaped somewhat like a horseshoe when viewed from above. Certainly there were clams in the environment of *Brachychampsa,* but the more likely prey were animals that were

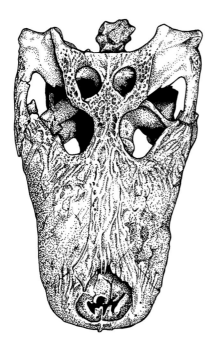

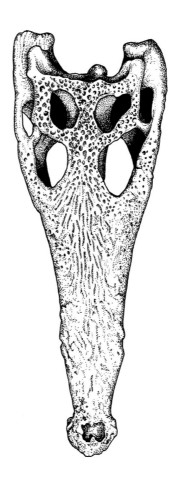

FIGURE 2.43
Skulls of *Brachychampsa* and *Leidyosuchus.*
Note the blunt, alligator-type skull of
Brachychampsa (after Brochu 1999), left, and
the long, tapered snout typical of crocodiles
in *Leidyosuchus,* right (after Schmidt 1938).

chased rather than dug for: turtles. In the Late Cretaceous turtles had become very abundant and were probably a rich food source—and where there is a food source, eventually a creature evolves to utilize it. Tooth marks on fossil turtle shells are common, and some Late Cretaceous turtle shell fragments have been found etched by (presumably) stomach acids. It is probable that many of the stomachs that etched these shells were those of *Brachychampsa* (Carpenter and Lindsey 1980).

When viewed from above, *Leidyosuchus,* in contrast, has a rather elongate, V-shaped snout lined with teeth and with a bulge at the end (see fig. 2.43). Fossil ichthyosaurs and mosasaurs, as well as modern porpoise and barracuda, have similarly shaped skulls, and all of these creatures are fish-eaters, so we can assume that *Leidyosuchus* mostly fed on fish. The teeth of *Leidyosuchus* are somewhat compressed and have sharp keels that are somewhat inwardly directed (Schmidt 1938)—perfect for catching fish. It was also a tooth (LACM-101165/28999) that proved that *Leidyosuchus* was in Baja California.

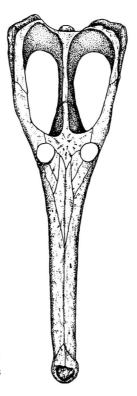

FIGURE 2.44
Skull of marine crocodile of the type
discovered in Oregon. After Andrews
1910–1913, vol. 2.

Morris, while working in Baja in 1973, found a "truly amazing" specimen, "the most complete and undistorted specimen of a Cretaceous crocodilian ever to have been reported" (pers. comm. 1999). It had characteristics of both alligators and crocodiles and may be an early relative of the modern alligators. Today it resides in the Baja California collection at the University of Mexico; it has not yet been described in a scientific paper.

Whether the crocodilians of Baja were all freshwater creatures or whether some dwelt in saltwater, as a few do today, we cannot tell. The lack of any crocodilian remains from marine deposits in Alta or Baja California may suggest that they were freshwater creatures, or that crocodilians didn't roam as far north in the Cretaceous fresh or saltwater environments of Alta California.

Older crocodilians have been found in an early Middle Jurassic limestone in central Oregon near the town of Suplee (Buffetaut 1979). Several vertebrae and skull fragments (see fig. 2.44) plus a broken limb bone suggest that a crocodilian from the family Teleosauridae (marine crocodiles) lived in the waters along the coast. This was the first example from the family to be found in North America.

The Oregon fossils, considered together with the evidence from Baja California, suggest that crocodilians should have been present in Alta California as well, at least during the mid-Jurassic, because California is farther south and hence presumably more tropical than Oregon during this time. The paucity of Mesozoic crocodilian remains in Alta California remains somewhat of an enigma.

Although we have surmised diets of turtles and fish for the ancient crocodilians of Oregon and Baja, perhaps like living forms they were opportunists. A misstep into the water could have led to disaster for a young or small dinosaur, or other vertebrates such as pterosaurs, mammals, and birds. Crocodilians may also have been a threat to marine reptiles if conditions were right.

The rich forested environment in California and Baja contained not only dinosaurs and crocodilians but also many other vertebrates. Richard Estes of San Diego State University reported a teid lizard (similar to the present-day *Paraglyphanodon*) and a small, crushing-toothed lizard of unknown affinities (Morris 1968b, 1974b) from Baja. Also found here are at least one terrestrial bird (see chapter 3) and even small mammals (Lillegraven 1970). These mammals of Cretaceous times were probably largely ignored by the dinosaurs, just as we hardly notice lizards scurrying out of our way. The only dinosaurs that paid them any attention at all would have been the smaller carnivorous types that ate them.

3

THE FLYING REPTILES

FLIGHT IS A COMPLICATED FUNCTION that took life literally billions of years to achieve. Insects were the first masters of the sky, evolving wings nearly 400 million years ago. It was not until the Mesozoic that reptiles achieved flight, and it took until the Cenozoic before mammals (bats) took to the air. Although today we have flying fish, flying squirrels, and even flying lizards, these are not true flyers but, rather, sophisticated gliders. Very specialized bones, muscles, and respiratory and circulatory sophistication are required to achieve the incredible coordination and stamina necessary for sustained flight.

Fossils of Mesozoic flying reptiles—the pterosaurs and the birds—are extremely rare, especially from coastal western North America. In order to fly, these creatures had to have exquisitely thin and often hollow bones, which are fragile and thus not likely to be preserved as fossils. Those that did become embedded in rock are again easily broken as they weather out of the substrate, or they may be simply so small that they are easily overlooked.

PTEROSAURS

Unlike birds that use feathers as flight surfaces, pterosaurs solved the problem of flight by having a long extended finger attached to a broad expanse of skin that trailed back along its body in the rear. Pterosaur fossils have been found primarily in marine sedimentary rocks, perhaps indicating that many pterosaurs were like pelicans and other ocean birds in lifestyle. But remains have also been found in floodplain deposits, and some of those far from shore. This perhaps shows a diversity of living habits or migrations of some species to inland areas during storms or breeding seasons. Bell and Padian (1995) report possible evidence of an Early

FIGURE 3.1

In this Late Cretaceous scene *Pteranodon* (a flying reptile) scoops up
fish, while a toothed bird, *Ichthyornis,* and a modern bird fly above.

Cretaceous pterosaur nesting site in a desert far inland from the South American shore, where there may have been less predation pressure.

The remains of three pterosaurs have been found in California, two from Butte County and one from Shasta County. In addition, Downs (1968) reported pterosaur remains from the Late Cretaceous Moreno Formation of the Panoche Hills in the western San Joaquin Valley. (This fossil was reported to have been deposited in the collection of Caltech [now at LACM]; however, no trace of this fossil has been found.) Also, a left humerus and two dorsal vertebrae of a pterosaur were found in marine Late Cretaceous rocks of eastern Oregon (Gilmore 1928). Thought first to be of a *Pteranodon*-like pterosaur, these fragments probably represent a species close to *Quetzalcoatlus* and other azhdarchid pterosaurs (Padian 1984). The presence of Late Cretaceous pterosaur remains in Oregon and northern California suggests these creatures may have been widespread in this region.

The Shasta County pterosaur was found by Patrick Embree in the Budden Canyon Formation of the Great Valley Group (Hilton et al. 1999). This find, a portion of a long bone (SC-VRF8) from a small concretion, is the oldest evidence of a flying vertebrate in the state. Ammonites (fossil mollusks) collected nearby establish an age for the pterosaur as Early Cretaceous, about 115 million years old (Murphy et al. 1969). Unfortunately, the fragment is not very informative as to the type of pterosaur it was from, as it is simply the midsection of a wing or leg bone (Padian, pers. comm., 1998). Fossil plants found in nearby strata indicate the animal may have lived near a warm, moist forest (Axelrod and Embree, pers. comm., 1995), probably in the foothills of the ancestral Klamath Mountains (Dickinson 1976; Nilsen 1986). Whether the pterosaur perished at sea, along a beach, or washed down a river we may never know.

The two Butte County pterosaur bones were found in 1998 by Eric Göhre at separate sites in the marine Late Cretaceous Chico Formation of Butte County. These bones, a fourth metacarpal (SC-VRF5) and an ulna (SC-VRF6), are from the wings of two large pterosaurs, and were verified with the help of Kevin Padian of the UCMP and Wann Langston of the University of Texas, Austin. Both bones were found in turbidites containing mollusk shells, shell fragments, bony fish teeth, and shark teeth, plus well-rounded cobbles of quartz and metamorphic rocks. The cobbles and some of the mollusk remains hint at a nearshore provenance for the material making up the turbidite. The well-rounded cobbles were formed in either a beach or river environment.

The largest of Göhre's pterosaur bones is the seventeen-inch fourth metacarpal. It is missing most of both articulating ends, and so would have been at least an inch or two longer. Complete, this metacarpal scales out to an animal with an approximate wingspan of at least fifteen to eighteen feet. Possible candidates include *Quetzalcoatlus* or *Pteranodon*, both found

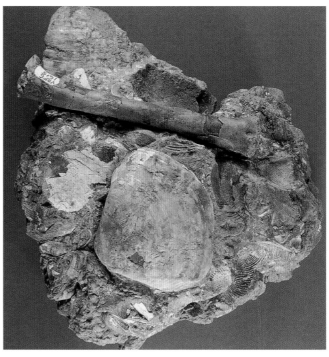

FIGURE 3.2

Fourth metacarpal (left) and ulna (right) of two large pterosaurs found in Butte County by Eric Göhre. Photos by the author.

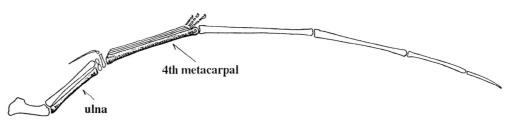

FIGURE 3.3

Position of fourth metacarpal and ulna in a pterosaur's wing.

in rocks of Late Cretaceous age in North America and both reaching sizes consistent with the length of this bone (Wellnhofer 1991). Comparison of this bone with the collection at the J. J. Pickle Research Campus at the University of Texas (with the help of Wann Langston Jr.) seems to preclude *Quetzalcoatlus,* as a slight bend in the metacarpal indicative of *Quetzalcoatlus* was absent; *Pteranodon* thus seems the more likely identity. *Pteranodon* was a toothless, crested pterosaur, the largest species of which, *P. sternbergi,* had a wingspan exceeding

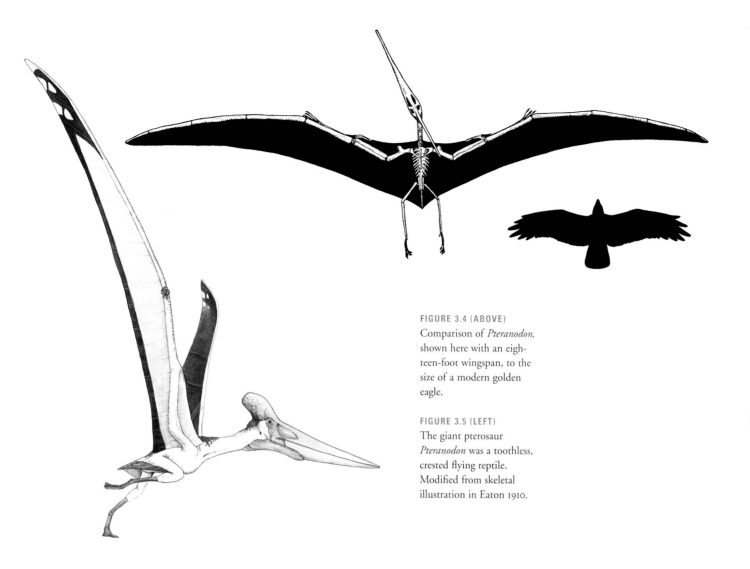

FIGURE 3.4 (ABOVE)
Comparison of *Pteranodon,*
shown here with an eigh-
teen-foot wingspan, to the
size of a modern golden
eagle.

FIGURE 3.5 (LEFT)
The giant pterosaur
Pteranodon was a toothless,
crested flying reptile.
Modified from skeletal
illustration in Eaton 1910.

thirty feet. The skull alone of *P. ingens* was nearly six feet long; it had relatively small eye
sockets and a short neck, but its long wings enabled it to soar far from the shoreline in a
probable search for fish (Wellnhofer 1991).

The second bone, the ulna, may be from a smaller animal than the metacarpal, or it may
be from a juvenile of the same species as the larger animal. According to Langston, the ulna
would scale up to an animal with about a six-foot wingspan.

It is interesting that both of these bones, though extremely fragile, managed to survive.
The metacarpal and ulna of pterosaurs, like all their long bones, are hollow and extremely
thin; flowing in the turbidity currents, they must have behaved like giant soda straws tum-
bling in the sediment-filled water. They survived being completely crushed only because

sediment was able to flow into the broken ends, filling the open tubes. Many pterosaur bones from North America have been found crushed by the weight of the overlying sediment (Gilmore 1928).

Fossil ferns, redwoods, and leafy flowering trees found by Göhre in the Chico Formation in Butte County indicate that the area where these animals lived was just offshore from a lushly forested ancestral Sierra Nevada.

BIRDS

It now seems fairly clear that birds evolved from small meat-eating dinosaurs and are the only close relatives of dinosaurs to have survived the mass extinctions at the end of the Mesozoic. Besides having many skeletal characteristics similar to dinosaurs, birds also have scales on their legs, and a few even have claws on their wings (hoatzin and screamers, for example). Some fossil birds had teeth as well. One Late Cretaceous bird found in Madagascar has a mosaic of characteristics of both birds and theropod dinosaurs (Forster et al. 1998), and the recent find of a bird-sized theropod in Early Cretaceous rocks in the Liaoning region of China proves that some dinosaurs were small enough to have evolved into birds (Xu et al. 2000). In the same deposits an eagle-sized dromaeosaur called *Sinornithosaurus millenii* was found, complete with downy featherlike structures, and although it did not fly (and was possibly secondarily flightless, as are ostriches), it had evolved the prerequisites for powered flight in its shoulder girdle. It is the most birdlike of all of the dinosaurs discovered thus far (Xu, Wang, and Wu 1999). Other dinosaurs (*Caudipteryx* and *Protarchaeopteryx,* for example) found in the same area suggest that featherlike structures may have had a broad distribution on theropod dinosaurs (Xu, Tang, and Wang 1999). A nonavian dromaeosaurid dinosaur from the Early Cretaceous has been found in China that has feathers identical to those found on modern birds, perhaps indicating that feathers evolved on dinosaurs before the emergence of birds (Norell et al. 2002).

The remains of a few Mesozoic birds have been found in Alta and Baja California. These specimens probably represent only a small fraction of the avifauna that existed in California during the Cretaceous.

The first such remains (LACM-33213) in western coastal North America were found in 1971 by H. J. Garbani and J. Loewe in the Late Cretaceous La Bocana Roja Formation of northern Baja California (Morris 1974c; Brodkorb 1976). They consisted of the following bones: from the shoulder, a left scapula and left coracoid; from the wing, the right ulna (a forearm bone); from the legs, a left femur and the distal end of a right femur (upper leg bones), and the right tibiotarsus (lower leg bone) (Brodkorb 1976).

FIGURE 3.6
Alexornis antecedens. Its bones were the first Mesozoic bird remains to be found in western coastal North America.

Brodkorb (1976) classified these sparrow-sized remains as belonging to a new genus and species of land bird he called *Alexornis antecedens.* Brodkorb named the genus *Alexornis* for his friend Alexander Wetmore; the species name means "going before in rank or time, ancestral," in reference to the supposed ancestry of this bird to the orders Piciformes and Coraciiformes.

In 1998 Eric Göhre found fossil bones from two species of birds at two sites in the Chico Formation of Butte County, California—the first evidence of Mesozoic birds found in Alta California. These bones, which were identified by Thomas Stidham of UCMP (Hilton et al. 1999), consist of a partial humerus (UCMP-170785) and a nearly complete ulna (UCMP-171185). The humerus is from the toothed, tern-sized bird *Ichthyornis,* which, much like living terns, is thought to have been a fish-eating plunge diver. The bone was found in a turbidite containing mollusk shells and fragments, bony fish teeth, and shark teeth. These materials probably had their origins nearshore and were later washed by turbidity currents into deeper water.

The ulna Göhre found is from a neognath, one of the earliest modern birds ever found, about the size of a modern pigeon. It was found in a small lens composed of shell hash also carried from a nearshore environment by a turbidity current.

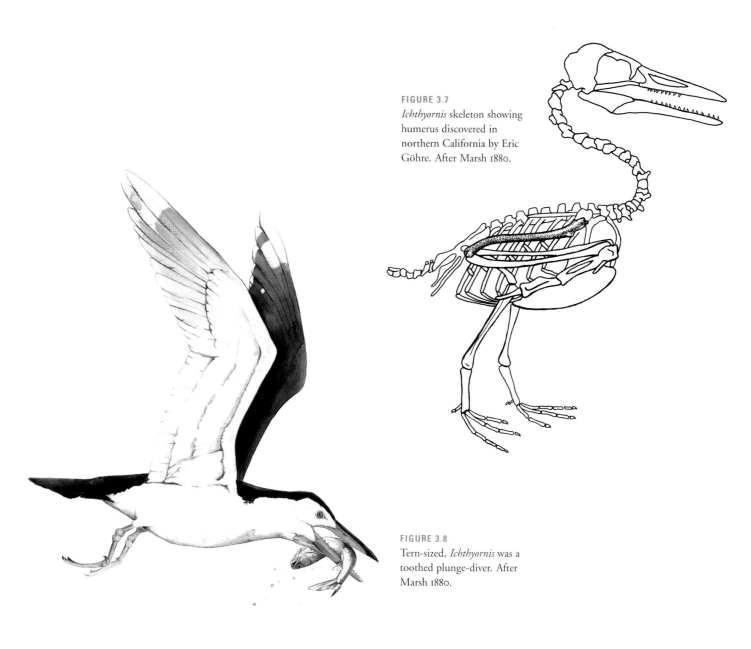

FIGURE 3.7
Ichthyornis skeleton showing humerus discovered in northern California by Eric Göhre. After Marsh 1880.

FIGURE 3.8
Tern-sized, *Ichthyornis* was a toothed plunge-diver. After Marsh 1880.

FIGURE 3.9
Position of ulna found by Eric Göhre in wing of modern bird.

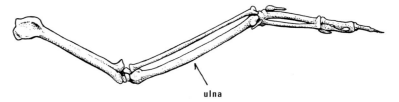

ulna

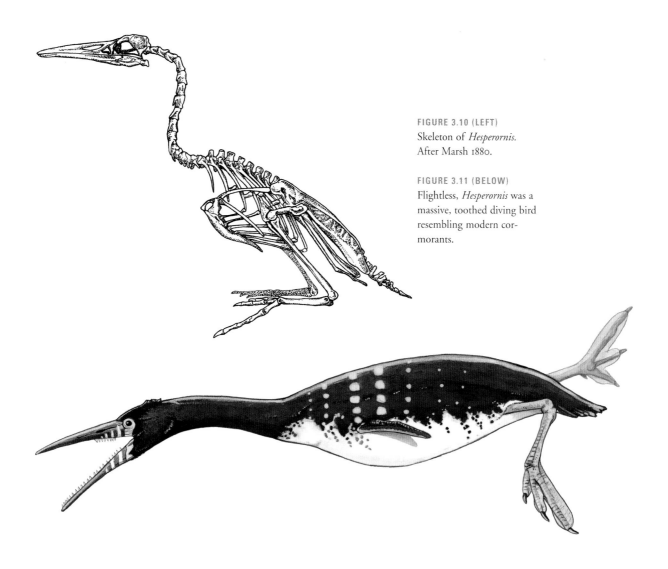

FIGURE 3.10 (LEFT)
Skeleton of *Hesperornis.*
After Marsh 1880.

FIGURE 3.11 (BELOW)
Flightless, *Hesperornis* was a
massive, toothed diving bird
resembling modern cor-
morants.

In 2000 Göhre found another bird bone in the same area, an extremely large phalanx (toe
bone) (SC-VBHE1). It was examined by Kevin Padian, who identified it as belonging to the
genus *Hesperornis,* based on comparison with material illustrated by Marsh (1880). *Hesperornis* was a large, flightless diving bird that is known from Late Cretaceous rocks from the
seaway that invaded central North America. The construction of its breastbone and shoulder shows that it evolved from a flying ancestor. Some specimens were over four feet long
with strong hind limbs and lobed feet, and they used their teeth to capture swimming prey
(Marsh 1880).

4

THE MARINE REPTILES

ON THE GALÁPAGOS ISLANDS, six hundred miles off the coast of Ecuador, land iguanas that probably rafted on masses of vegetation to the sparsely vegetated island shores turned to the sea for sustenance. They evolved into marine iguanas (see fig. 4.1). Today they are often found basking in the sun to warm their bodies; after sufficient warming they plod to the ocean and, with legs folded back, undulate their body and flattened tail to swim out from shore. Diving through the cool salt water, they graze the bottom for seaweed, spending half an hour or more before returning to shore to rewarm their bodies. Like sea turtles, marine iguanas lay their eggs on shore.

FIGURE 4.1
Marine iguana of the Galápagos Islands, Ecuador. Photo by the author.

A similar process of adaptation to a marine life apparently occurred with other reptiles in the past, and in some cases these animals adapted to a permanent oceanic existence. Some of these ancient marine reptiles may have remained herbivores like the marine iguana, but most developed the swiftness necessary for catching prey. The less evolved forms may have returned to the beaches to lay eggs, but more advanced forms simply gave birth at sea. Land-dwelling reptiles have been lured to water environments as far back as the Paleozoic. *Hovasaurus,* an aquatic reptile from the Late Permian (ca. 250 MYBP), may have used undulations of its body and tail for swimming (Benton 1998), not unlike the marine iguanas of the Galápagos Islands.

79

THALATTOSAURS

Thalattosaurs were marine reptiles that thrived during the Triassic. *Thalattosaurus* means "marine lizard." Thalattosaurs are known from predominately fragmentary remains and are here described from information gleaned by Nicholls (1999) in her description of *Thalattosaurus alexandrae.*

At a distance a thalattosaur probably resembled a crocodile or large Nile monitor lizard when using its claws to crawl out upon a reef (Nicholls 1999). Upon closer inspection, however, its head would look dramatically different from other reptiles, having a long snout with sharp, grasping teeth at the front of its lower jaw and rasping hooked teeth in the opposing skull. Most likely carnivorous, thalattosaurs were probably excellent swimmers and used their long and flattened, eel-like tails in concert with undulations of their bodies for propulsion. Their limbs were most likely webbed and paddlelike, with well-developed grasping claws on the fingers and toes (see fig. 4.2).

The California thalattosaur remains come from the Late Triassic Hosselkus Limestone in Shasta County, where more than two hundred Triassic fossil marine reptile discoveries have been made. The first fossils of thalattosaurs were discovered in 1893 by James Perrin Smith of Stanford University, and more discoveries soon followed. These fossils were studied in the early 1900s by the University of California, Berkeley, paleontologist John C. Merriam. Although some of his designations for these creatures were later revised, much of his pioneering work has survived the test of time.

Merriam established the order Thalattosauria, which at the time included two genera, *Thalattosaurus* Merriam (1904) and *Nectosaurus* Merriam (1905b). Thalattosaurids have been found in Triassic rocks in North America, Europe, and recently in China (Rieppel et al. 2000). Recent work by Elizabeth Nicholls (1999) of the Royal Tyrrell Museum of Alberta, Canada, has clarified much of the taxonomy of the California thalattosaurids.

Thalattosaurus alexandrae (UCMP-9084) was named by Merriam in 1904 in honor of Annie Alexander, a skilled fossil hunter and patron to the University of California Museum of Paleontology. *Thalattosaurus alexandrae* was about six feet long and an excellent swimmer (Nicholls 1999). Similar to primitive ichthyosaurs, *T. alexandrae* probably had a long and flattened, eel-like tail. It may have used its claws to withstand the force of the surf when it crawled up on shore, much like the marine iguanas of the Galápagos. Whether it frequently retreated to the shore to warm its body or mainly just to lay eggs we may never know.

Behind its sharp, grasping teeth and rasping hooked snout, *T. alexandrae* had, in both the skull and jaws, toothlike conical knobs that were probably used more for holding and crushing than actual chewing. Its most likely prey were swimming mollusks such as am-

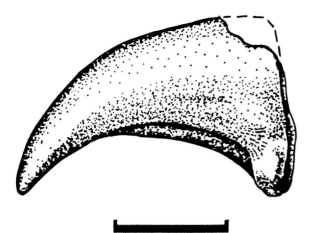

FIGURE 4.2 (LEFT)
Claw of *Thalattosaurus alexandrae*. Scale bar equals 2 cm. From Nicholls 1999; courtesy of the University of California Museum of Paleontology.

FIGURE 4.3 (BELOW)
Skull of *Thalattosaurus alexandrae*. After Nicholls 1999.

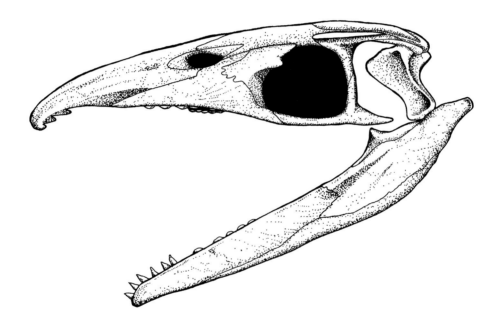

monoids, which are commonly found as fossils in the same rocks as *T. alexandrae* (Nicholls 1999). These shelled creatures, related to squid and octopus, were similar to the chambered nautilus swimming in tropical seas today. The nautilus has a fleshy head with well-developed eyes and a beak in the middle of grasping tentacles that are used for capturing its prey. All of this is tucked into a chambered shell. The chambers are used to control buoyancy at different levels in the sea, much like modern submarines use gas in chambers to maintain or change depth. The flesh of the animal occupies only the last chamber, called the living chamber.

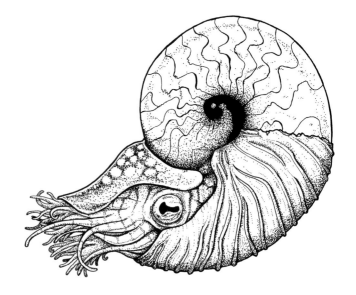

FIGURE 4.4 (RIGHT)
Ammonoid in chambered shell.
Shown with exposed inner chambers.

FIGURE 4.5 (BELOW)
Hypothetical skeleton of a
Thalattosaurus. Based partly on
illustrations by Sloan in Nicholls
1994, 1999; postcranial skeleton
based on *Askeptosaurus* in Kuhn-
Schnyder 1974.

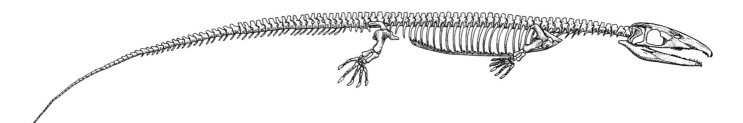

Although we lack complete soft parts for ammonoids, we presume that they were very much like those of the chambered nautilus.

Merriam described two other species from the genus *Thalattosaurus,* but it is likely that neither belong in this genus. Although the skull of the type specimen of *Thalattosaurus perrini* (named for James Perrin Smith of Stanford University in 1905) cannot be located, Nicholls (1999) says that old photographs of the specimen indicate that it is clearly not a thalattosaur. *Thalattosaurus shastensis* (UCMP-9120, described by Merriam in 1905b) is, according to Nicholls (1999), most likely *Nectosaurus.*

Nectosaurus halius (UCMP-9124) was first described by Merriam (1905b); *Nectosaurus* means "swimming reptile," and *halius,* "belonging to the sea." This animal was about three feet long, approximately the size of an iguana. Its thin teeth were conical and pointed, with finely incised parallel lines on their sides, and all were set in deep sockets. These teeth suggest that *N. halius* was a carnivore, perhaps feeding on fish or other soft-bodied swimmers.

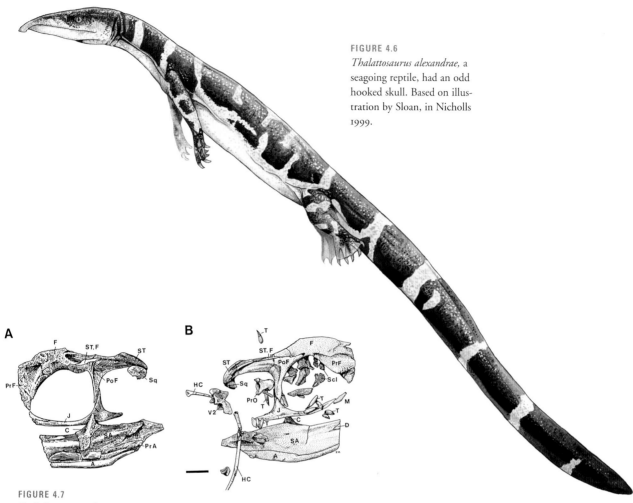

FIGURE 4.6
Thalattosaurus alexandrae, a seagoing reptile, had an odd hooked skull. Based on illustration by Sloan, in Nicholls 1999.

A

B

FIGURE 4.7

Skull fragments of *Nectosaurus* sp. (a = medial view, b = lateral view). Scale bar equals 2 cm. From Nicholls 1999; courtesy of the University of California Museum of Paleontology.

Although known only from partial skeletal material, *N. halius* apparently lacked both the crushing teeth and the more stout, conical teeth of *Thalattosaurus.* So it probably did not eat the shelled mollusks that *Thalattosaurus* may have consumed. The identifiable specimens of *Nectosaurus halius* are all small, and it is possible the type specimen is that of a juvenile. Associated unidentifiable bones, some four times as large, suggest that *N. halius* may have grown much larger. Another interpretation could be that the larger bones came from an as yet unidentified larger species (Nicholls 1999).

ICHTHYOSAURS

Ichthyosaurs were highly evolved, streamlined reptiles that looked more like fish than reptiles. The name ichthyosaur literally means "fish reptile" in Greek. Like modern whales, their fishlike shape evolved for swift swimming for the purpose of catching prey. Most were dolphin-sized or smaller, but some were the size of modern whales, reaching lengths of sixty feet or more (Dupras 1988). A recent discovery of a seventy-foot ichthyosaur in British Columbia is the largest marine reptile ever found (Nicholls, pers. comm. 2000). Its skull alone is nearly nineteen feet long (Grady 2001). Ichthyosaurs swam the Mesozoic seas from 245 to 90 million years ago, and their fossils have been found on every continent except Africa and Antarctica. According to Larry D. Martin of the University of Kansas at Lawrence (pers. comm., 2002), the diversity of ichthyosaurs waxed and waned according to the temperature of the Earth. When the climate was warm the number of species rose, and when it was cooler fewer species survived.

Until recently the origin of ichthyosaurs could only be guessed at. Scientists were fairly confident that they evolved from a smaller land-dwelling reptile that moved back to the sea to fill predatory niches, much as ancestral dolphins evolved from four-legged, land-based mammals. Our ideas were confirmed by two fossils of early ichthyosaur relatives, *Utatsusaurus* and *Chaohusaurus,* that were recently discovered in Asia (Motani 2000). Both had two pairs of limbs and looked much like lizards with flippers. Careful analysis of *Utatsusaurus hataii* indicates that the ichthyosaurs are diapsids, but they are not included with the Sauria, the group that contains the lizards, crocodiles, and birds (Motani et al. 1998).

Early ichthyosaurs had eel-like body proportions and lacked a dorsal fin and fishlike tail. Although most Late Triassic California ichthyosaurs retained an eel-like tail, one more advanced species, *Californosaurus perrini,* begins to show the start of a tail bend, which would evolve toward the more fishlike tail used for swifter locomotion (Motani, pers. comm. 1999). Motani estimates that a six-foot ichthyosaur could have swum as fast as a tuna but perhaps a bit slower than the fastest modern whales (Stokstad 2000).

The front paddles of ichthyosaurs evolved by losing the thumb and adding more digits and wrist bones. The bones became wider, flatter, and more closely packed, and, in contrast to land-based vertebrates, the wrist bones became indistinguishable from one another. The flipper surface became a relatively rigid, flat hydrodynamic panel similar to the flipper of a modern whale. Paired paddles fore and aft were probably kept extended and utilized like hydroplanes for accurate steering and quick turns while feeding and possibly during mating. They could also be used for ascending and descending (McGowan 1983b).

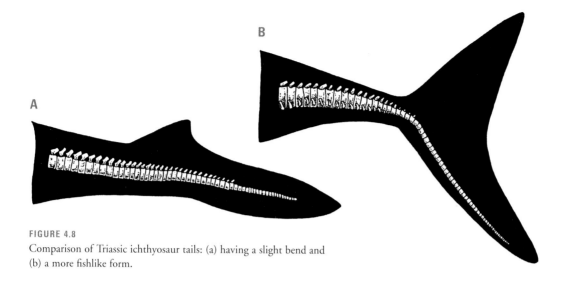

FIGURE 4.8
Comparison of Triassic ichthyosaur tails: (a) having a slight bend and (b) a more fishlike form.

As the ichthyosaur body became larger and more streamlined, the vertebrae changed from the shape of a soup can to that of a hockey puck. This change allowed the body to be stiffer and the tail to become the means of locomotion. Now the animal could efficiently cruise the open ocean and dive into its depths for squidlike prey (Motani 2000).

Some ichthyosaur fossils from Germany and other parts of the world preserved not only the bones but also some soft tissue, suggesting outlines of their bodies. The dorsal fin and true shape of the tail on some ichthyosaurs would not be known but for these beautifully preserved outlines in a carbonaceous film (see fig. 4.9). Most ichthyosaurs probably had a dorsal fin that acted much like the flight feathers on an arrow, keeping the animal from rolling while it swam through the water.

Ichthyosaurs were efficient predators, most swimming fast enough to catch fish and squid. According to Benjamin P. Kear, of the South Australia Museum at Adelaide, even the remains of hatchling turtles have been found in one individual (pers. comm., 2002). Also included in the diet of some were nautiloids, belemnites, and ammonoids (the shelled relatives of modern squid and octopus). These cephalopods were abundant in Mesozoic seas. Small hooklets that were once part of the suckers on cephalopods, as well as fish scales, have been found as stomach contents of some fossil ichthyosaurs (McGowan 1983b). Recent evidence presented by Peter Doyle and Jason Wood (Holden 2002) suggests that some Jurassic ichthyosaurs that ate belemnites (internal skeletons of a squidlike animal) may have regurgitated the skeletal portions, much as whales regurgitate squid beaks and owls regurgitate the

FIGURE 4.9
Carbonaceous film on the ichthyosaur *Stenopterygius quadricissus*
(CSGEO 72728) found in Jurassic rocks in Germany reveal the outline
of the body and placement of the dorsal fin and tail. Courtesy of the
Field Museum, Chicago.

fur and bones of rodents. Some of the thick, calcium carbonate skeletons of the belemnites
appear to have been etched, presumably by the stomach acid of the ichthyosaur.

Like modern dolphins, most ichthyosaurs had many sharp, peglike teeth mounted in long,
pointed jaws that were perfect for grasping slippery prey. Their teeth were so precisely spaced
that they fit together alternately on the top and bottom jaws, leaving no space between them.
The jaw muscles were set quite far back so that a quick-closing action resulted in an instan-
taneous snap of their tooth-lined jaws (McGowan 1983b).

Ichthyosaurs also had large eyes framed by a bony eye ring that looked much like the
iris on an expensive camera lens. Keen eyesight enabled them to see in dimly lit waters and
was probably their primary means of locating food. The largest ichthyosaurs had eyes five
times the diameter of those of an African elephant (Motani 2000). Evidence in the braincase and fossil ear structures suggests that ichthyosaurs did not use echolocation (reflective
sound) as a way to find food, as modern whales and dolphins do (McGowan 1983b). However, recent evidence from the braincase and from blood and nerve canals that service the
rostrum (beak) suggests that some species may have had sensitive skin on the face and jaws,
which may have enabled them to sense prey better. One species has been found to have
thin, convolute bony nasal structures, possibly to make smell more acute (Kerr, pers. comm.
2002).

Like mammals, reptiles must breathe air, so ichthyosaurs had to rise to the surface to
breathe. Sea turtles crawl out onto the beaches to lay their eggs, but ichthyosaur fossils show
that they bore their young like whales, alive in the water and ready to swim (McGowan 1983b).

Ichthyosaurs became extinct in the Late Cretaceous, about 90 million years ago, 25 mil-

FIGURE 4.10
"Live birth" as shown in fossil ichthyosaur *Stenopterygius* (SMNS6293)
that possibly died while giving birth. Courtesy of Rupert Wild and the
Museum of Natural History, Stuttgart, Germany.

lion years before the mass extinctions at the end of the Mesozoic. Bardet (1992) conjectures
that they may have become extinct because of the decline and extinction of prey species such
as belemnites. Motani (2000) notes that at about the time that ichthyosaurs disappeared,
more advanced sharks evolved. It is interesting to note also that at about this same time
mosasaurs evolved from varanid lizards and took to the sea to exploit open niches. Like the
sharks, many mosasaurs probably frequented a nearshore environment, while the larger advanced ichthyosaurs likely exploited a more open ocean habitat. All of these occurrences may
be coincidental, but perhaps we should be looking for a change in the ocean environment
at this time that could have driven these extinctions and evolutionary changes.

Triassic Ichthyosaurs

Triassic ichthyosaurs were widely diversified and highly specialized animals (McGowan 1983b),
filling niches similar to those of whales, seals, and dolphins today. Like these mammals, the
ichthyosaurs probably evolved to these niches because of the largely untapped supply of fishes
and cephalopods. Triassic ichthyosaurs of the world, especially the shastasaurids, are associated with reefal limestones, including most notably the Late Triassic Hosselkus Limestone
of Shasta County, California. This may mean that these forms were adept at ambush types
of predation associated with complex reef structures—excellent environments for surprise
attacks (Massare and Callaway 1987).

A recent work by Ryosuke Motani (1999), then at the University of California Museum
of Paleontology in Berkeley, has done much to organize the ichthyosaurs according to modern cladistic methods. Because of numerous name changes and classification modifications,

it is important here to update the classification of the Triassic reptiles of California to understand the historical literature. In the new classification, California ichthyosaurs from the Triassic fall into two families: Shastasauria and Euichthyosauria.

SHASTASAURIA   The family Shastasauridae is known from California, Oregon, Nevada, Mexico, and Canada as well as possibly Alaska. Outside North America it is also known from China, the former USSR, Svalbard, western and eastern Europe, and New Caledonia (Callaway and Massare 1989b). In California, the family comprises three genera: *Shastasaurus, Californosaurus,* and *Shonisaurus.*

Remains from several members of the genus *Shastasaurus* (Merriam 1895) have been found in California. Nicholls (2000), however, reports that the genus *Shastasaurus* may apply only to Merriam's type specimens of *S. alexandrae* and *S. osmonti* (UCMP-9076; reclassified as *S. pacificus* by Motani 1999). *Shastasaurus* were medium- to large-sized ichthyosaurs that usually reached lengths of thirteen to sixteen feet, although some may have reached up to thirty-nine feet (McGowan 1983a; Callaway and Massare 1989a). The teeth were sharp and peglike and roughly of equal size, much like those of some dolphins.

One *Shastasaurus* specimen, found by Merriam, has not been assigned a species. A composite restoration of the skeleton by Callaway and Massare (1989a) depicts this ichthyosaur with a slight tail bend and sharply tapering skull. All four flippers were about the same size and fairly narrow.

Shastasaurus pacificus (UCMP-9077), meaning "Shasta reptile of the Pacific," was first described by John C. Merriam in 1895. Today it is the only shastasaur in California to survive reclassification. *Shastasaurus osmonti* (Merriam 1902a) (UCMP-9076), *S. alexandrae* (Merriam 1902a) (UCMP-9017), and *S. altispinus* (Merriam 1902a) (UCMP-9083) have all been reclassified by Motani (1999) as *Shastasaurus pacificus.* Motani reclassified *S. careyi* Merriam 1902a (UCMP-9075) as *Shonisaurus* sp.

Shastasaurus pacificus was robustly built and twelve to fifteen feet long, about the size of a large dolphin (Dupras 1988). Although fossils are incomplete, both the fore and hind fins appear to have been narrow, with perhaps only three digits on each. It had more foreshortened vertebrae and robust neural spines than other examples of the genus (Sander in Callaway and Nicholls 1997).

The genus *Californosaurus* is represented by a specimen that was first called *Shastasaurus perrini* (Merriam 1902a) (UCMP-10998) and then later placed in its own genus, *Delphinosaurus* (Merriam 1905c). This genus was then replaced by Kuhn in 1934 by *Californosaurus* because the generic name *Delphinosaurus* was already taken. Today it is classified as *Cali-*

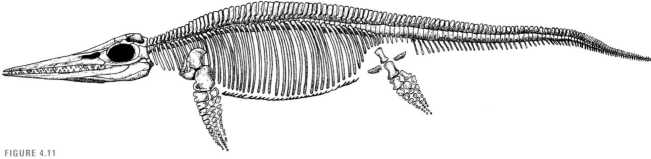

FIGURE 4.11
Reconstruction of *Shastasaurus altispinus* (now *pacificus*). From Callaway and Massare 1989b.

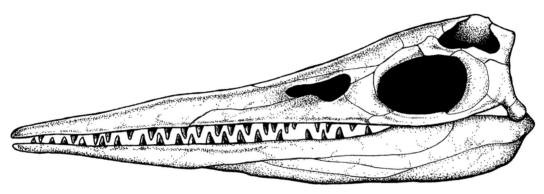

FIGURE 4.12
Skull of *Shastasaurus altispinus* (now *pacificus*). From Callaway and Massare 1989b.

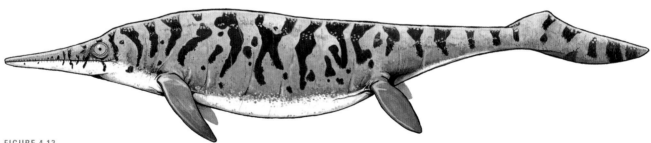

FIGURE 4.13
Shastasaurus pacificus was a fully marine reptile about the size of a large dolphin. After Callaway and Massare 1989b.

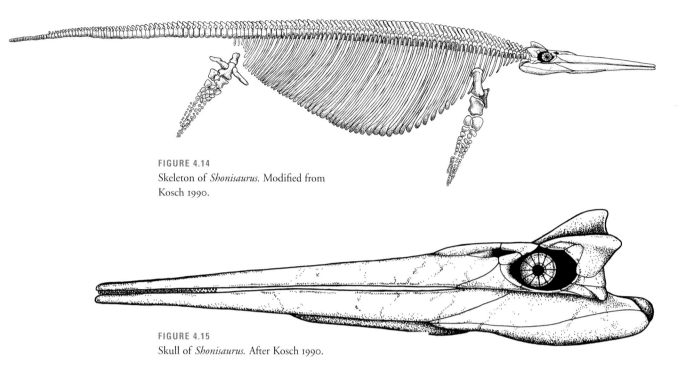

FIGURE 4.14
Skeleton of *Shonisaurus*. Modified from
Kosch 1990.

FIGURE 4.15
Skull of *Shonisaurus*. After Kosch 1990.

fornosaurus perrini (Motani 1999). The genus *Perrinosaurus* (Merriam 1934) has also been changed to *Californosaurus;* therefore *Perrinosaurus perrini* is now known as *Californosaurus perrini* as well.

Camp (1976) coined the name *Shonisaurus* to describe large, long-bodied (sometimes whale-sized) ichthyosaurs found as fossils in Late Triassic rocks in central Nevada. Named for the Shoshone Mountains, these are among the largest ichthyosaurs ever found. *Shonisaurus* was more advanced than *Mixosaurus* and *Cymbospondylus* (two Middle Triassic ichthyosaurs) but less specialized than Jurassic forms (Camp 1980). Camp (1980) recognized that *Shonisaurus* from Nevada closely resembled the Late Triassic ichthyosaur *Shastasaurus careyi,* found in California's Shasta County. Accordingly, Motani (1999) reclassified *Shastasaurus careyi* (UCMP-9075) as *Shonisaurus* sp. Slightly smaller than adult Nevada specimens, the Shasta County specimen may be a subadult.

The skeleton of *Shonisaurus* was robust with thick, heavy bones. Its slender fore and hind fins were about the same size, both about six feet long (McGowan 1983b). A slight tail bend suggests that it had a modest caudal (tail) fin. Although a complete skull has never been found, its crested skull had a long, thin rostrum, or beak. It appears to have had relatively small eyes set quite far back in the skull. The teeth are small and set in sockets, and appear to be re-

FIGURE 4.16 (LEFT)
Sierra College student Tom Coker stands in front of a plaque of *Shonisaurus* at Berlin-Ichthyosaur State Park, Nevada. Photo by the author.

FIGURE 4.17 (BELOW)
Shonisaurus is one of the largest ichthyosaurs ever found. Modified from Kosch 1990.

stricted to the front of the jaws. Except for *S. mulleri* (Camp 1976), the rib tips are expanded, unlike any other ichthyosaur found (McGowan 1983b).

EUICHTHYOSAURIA   The family Euichthyosauria, which means "true ichthyosaurs," comprises three genera in California: *Toretocnemus, Merriamia,* and *Californosaurus.*

Toretocnemus were small animals, only about three to four feet long (McGowan 1983b; Merriam 1908b). Both the fore and hind fins appear to have been narrow and similar in length (McGowan 1983b). First described by John C. Merriam (1903a), *Toretocnemus californicus* (UCMP-8100) was the only species described in the genus. Today, however, *Merriamia zitteli* Merriam 1904 (UCMP-8099) has been assigned to *Toretocnemus* by Motani (1999). Both species are known only from fragmentary material.

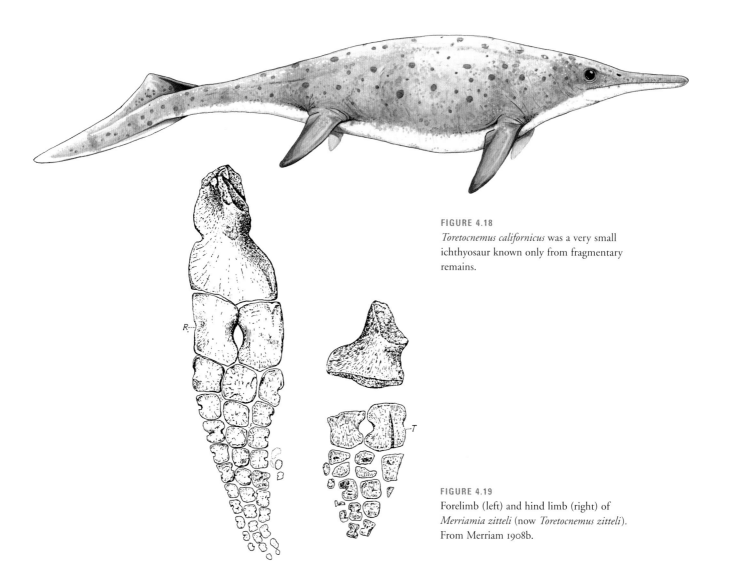

FIGURE 4.18
Toretocnemus californicus was a very small ichthyosaur known only from fragmentary remains.

FIGURE 4.19
Forelimb (left) and hind limb (right) of *Merriamia zitteli* (now *Toretocnemus zitteli*). From Merriam 1908b.

Both *T. californicus* (UCMP-8100) and *T. zitteli* share many characteristics, including those of their fins. *T. zitteli* was originally named *Merriamia zitteli* by Boulenger 1904 (UCMP-8099). Motani (1999) changed the genus to *Toretocnemus* because *T. zitteli* is very similar to *T. californicus;* the only difference may be that in *T. zitteli* the forefin is longer than the hind fin, while in *T. californicus* the fore and hind fins are nearly equal (see fig. 4.19).

Now, too, the genus *Merriamia* replaces the genus *Leptocheirus,* which was created by Merriam in 1903. Motani made this change because *Leptocheirus* had already been used to de-

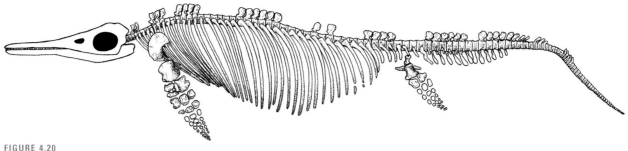

FIGURE 4.20

Californosaurus skeleton, from information provided by Motani. Skull and distal flipper bones are artist's reconstruction.

FIGURE 4.21

The dolphin-sized *Californosaurus perrini* was named for Stanford paleontologist James Perrin Smith. Based on reconstruction information provided by Motani.

scribe another animal (Motani 1999). Thus, *Leptocheirus zitteli* Merriam 1903a (UCMP-8099) becomes *Merriamia zitteli.*

The genus *Californosaurus,* proposed by Kuhn (1934), means "California reptile." The species *perrini* (UCMP-10998) is named for James Perrin Smith, the first to find fossil reptiles in Shasta County. It was a dolphin-sized animal reaching lengths of six to ten feet. It appears to have had narrow forefins and was the first known ichthyosaur to develop a downward bend in the tail (McGowan 1983a), which gave rise to more fishlike tails and ultimately the tuna-shaped tail. According to Motani (1999), *Shastasaurus perrini* Merriam 1902a, *Delphinosaurus perrini* Merriam 1905c (UCMP-9119), and *Perrinosaurus perrini* Merriam 1934 are all considered synonymous with *Californosaurus perrini.*

Jurassic Ichthyosaurs

By the Jurassic, California had developed a true coastline. No longer just a series of islands, the Jurassic shoreline (which would roughly slice the present-day state in half lengthwise) was a rugged coast sitting below a volcano-studded mountain range (see fig. 1.11). Evidence of marine reptiles from the Jurassic in California is rare; only fragmentary remains of plesiosaurs and ichthyosaurs have been found thus far.

Two ichthyosaur remains were found in the radiolarian cherts of the Franciscan Formation. Radiolarian cherts are formed from deepwater sediments consisting mostly of the nearly microscopic siliceous skeletons of radiolaria that settled onto the deep open ocean bottom (see fig. 1.17). These layered deposits were later scraped off the ocean crust where the ocean floor was being subducted beneath the continental margin (see fig. 1.8). Convergent squeezing normally obliterates larger fossils, but somehow these two fragments of ichthyosaur rostra survived. The rostrum is the long, pointed muzzle of the animal, containing the teeth of the skull and mandible. At the time that the information about the two rostra was published (Camp 1942b), most of the Franciscan was considered to be Late Jurassic. After careful analysis by Camp, the fossils seemed to be most closely related to Late Jurassic forms.

The fossils, however, were discovered not in the Coast Range, where these rocks are found in place, but in the San Joaquin Valley, where they had been transported by recent streams. One of them, originally named *Ichthyosaurus californicus,* was found in Stanislaus County, and the other, *Ichthyosaurus franciscus,* was found in San Joaquin County.

Ichthyosaurus franciscus (UCMP-33432), meaning "fish reptile of the Franciscan Formation," was described by C. L. Camp (1942b) from a water-worn chert cobble containing a portion of an ichthyosaur rostrum. The cobble had washed down a creek (Corral Hollow) from the Coast Range Province into the San Joaquin Valley where it was found. N. L. Taliaferro determined that the rock was a radiolarian chert by studying a thin section of it, and found it to be identical to the radiolarian cherts of the Coast Range Province. Other thin sections were studied by Arthur S. Campbell, who found an identifiable radiolarian fossil (Camp 1942b).

The structure of the rostrum and teeth were determined from cross sections of the rock. The simple teeth were fairly small, slightly curved cones and were closely spaced and numerous, set loosely in an open dental groove. The rostrum, tapered and moderately stout, was not from one of the slender-snouted ichthyosaurs (Camp 1942b). Camp compared the specimen with other species of ichthyosaur and decided that it was a new species. McGowan

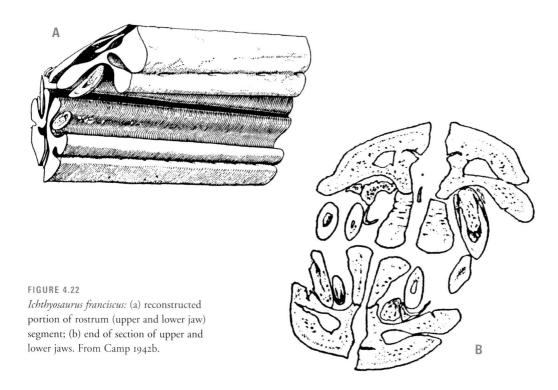

FIGURE 4.22
Ichthyosaurus franciscus: (a) reconstructed
portion of rostrum (upper and lower jaw)
segment; (b) end of section of upper and
lower jaws. From Camp 1942b.

(1976) concluded, however, that there is inadequate material to be able to reconstruct the whole animal, and hence to erect a new species.

Ichthyosaurus californicus (UCMP-36394), or "fish reptile of California," was described by Camp (1942b) from another water-worn chert cobble containing a portion of an ichthyosaur rostrum. This cobble also originated in the Coast Ranges Province and washed down a creek into the San Joaquin Valley. It was found near the mouth of Del Puerto Canyon, which cuts into the Great Valley Group rocks along the northwestern side of the valley. According to Camp (1942b), the specimen had six visible teeth on one side and consisted almost entirely of the conjoined dentaries and premaxillaries (lower and upper jaws). In the lab the rock containing the rostrum was sawed through and the ends were polished to enhance the detail for study.

The rostrum was similar to that of *Ichthyosaurus franciscus* but from a larger animal. The teeth were also much larger and more widely spaced than in *I. franciscus. Ichthyosaurus californicus* also had structural differences in the bones of the rostrum. These characteristics suggested to Camp that it was a new species. McGowan (1976) concluded, however, that there is inadequate material to be able to erect a new species.

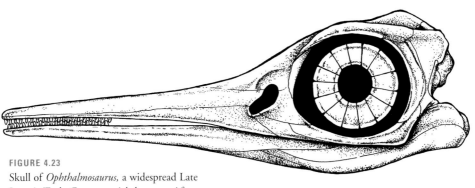

FIGURE 4.23
Skull of *Ophthalmosaurus*, a widespread Late
Jurassic/Early Cretaceous ichthyosaur. After
Andrews 1910.

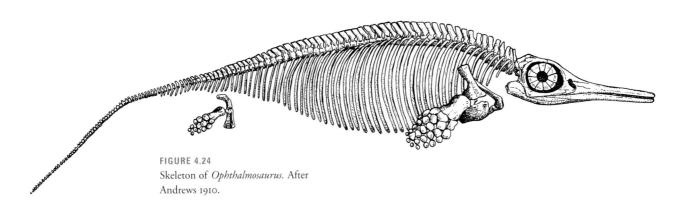

FIGURE 4.24
Skeleton of *Ophthalmosaurus*. After
Andrews 1910.

FIGURE 4.25
Ophthalmosaurus. Based on reconstruction
by Motani.

PLESIOSAURS

Several discoveries of plesiosaurs from the Jurassic and Cretaceous have been made in California. Plesiosaurs are some of the most interesting marine reptiles to have ever lived. With a body shaped like a somewhat flattened sweet potato, these huge animals had a relatively short tail and four flippers that resemble the flippers of a humpback whale. The flippers were not used as paddles for rowing but more as wing foils for virtually flying through the water like penguins. Each appendage could be used independently for quick maneuvering during chases, mating, or changing depth (Robertson 1975). Although some species were short-necked, many others had long, snakelike necks. The short-necked species may have relied on ambush and quick bursts of speed to overtake prey, while the long-necked species probably used their necks much like a striking snake to dart out at fast-moving prey (Taylor 1981). Their relatively small skulls were replete with long, sharp, curving teeth that protruded from both the upper and lower jaws. Efficient predators, they ate fish and other swimming prey such as nautilus, ammonites, squid, and perhaps even jellyfish.

Jurassic Plesiosaurs

Plesiosaur fossils from the Jurassic are known from the northwestern Sacramento Valley, San Luis Obispo County, the foothills of the central Sierra Nevada, and Orange County.

The Late Jurassic beds of the Great Valley Group of the northwestern Sacramento Valley have yielded the most Jurassic plesiosaur remains. The first mention of plesiosaurs in this area is in a letter by Merriam dated March 17, 1908, in which he mentions what he thought was a plesiosaur tooth from the Chico Formation. At that time, the term *Chico Formation* was used for most of the rocks that are today considered the Great Valley Group. Today, in contrast, the designation is restricted to the youngest portion—upper Cretaceous—of the Great Valley Group, exposed mainly on the western side of the Sacramento Valley. Where this fossil was found and whether it was from Jurassic or Cretaceous rocks seems to have been lost with the specimen.

From Merriam's letters it seems that in 1913 Thomas L. Knock, a civil engineer in Willows, found what are probably plesiosaur remains in Jurassic rocks south of the town of Elk Creek. He recounted in a letter to Merriam, "The vertebrae are about 5 inches in diameter, then there are a lot of small bone [*sic*], rib bone apparently. . . . The whole thing was some thirty or forty feet long." He enclosed a map of its location plus pictures of some of the bones. The pictures of the vertebrae suggest that they were indeed most likely from a plesiosaur. What happened to these fossils we may never know, but it is clear that Merriam, if he was not able to follow up and obtain this find, lost an important opportunity to collect one of the most complete Jurassic reptilian specimens ever found in California. The author has made a field search for other remnants of the specimen, so far without success. A possible mold of a long bone from a flipper was found in rocks at what appears to be the site.

Approximately twenty-five Late Jurassic plesiosaurian discoveries have been made in the hills along the western Sacramento Valley, but thus far only one specimen has been assigned a genus name. It is a cervical vertebra (UCMP-56204), found by Bates McKee of the University of Washington in the summer of 1960, and is listed in the Berkeley collection as *Dolichorhynchops (Trinacromerum),* from Lower Cretaceous rocks. On checking the coordinates for the site, however, it is more likely that it is from Late Jurassic rocks. Welles (unpub.) referred to this specimen as *Trinacromerum.* It is doubtful whether the genus can be identified from a single cervical vertebra, let alone as either *Dolichorhynchops* or *Trinacromerum.*

In any case, *Dolichorhynchops* was a short-necked plesiosaur first described by Williston (1902). He described *Dolichorhynchops osborni* from a nearly complete skeleton found in Kansas. It has a long and narrow, tooth-studded snout, and the overall shape of its skull is reminiscent of the Indian crocodilian called the gavial. Like the gavial, *Dolichorhynchops* was probably a fish eater. A cartilaginous pocket in the front of its nasal chamber may have housed an organ used to smell underwater prey. Whereas the name "pliosaur" referred to the true short-necked plesiosaurs, on the basis of skull morphology Carpenter places the somewhat longer-necked *Dolichorhynchops* in the family Polycotylidae, a sister group to the long-necked elasmosaurids (Carpenter, in Callaway and Nicholls 1997).

The most recent discovery of plesiosaur remains from the Sacramento Valley, consisting of several individual bones and a tooth, was made by volunteers from the Sierra College Natural History Museum, ranchers, the author, and his brother Paul Hilton. These fossils were found in latest Jurassic rocks of the Great Valley Group along the western side of the Sacramento Valley, in an area south of the town of Elk Creek, to the north of Paskenta. Glenn Storrs, curator of vertebrate paleontology at the Cincinnati Museum Center, has examined these remains and reports, "It appears that there is a short-necked form (pliosaurid of old usage) and what appears to be an elasmosaur-like animal." However, without more diagnostic material it will be difficult to attach a genus name to any of these specimens.

In 1949 two vertebrae from a plesiosaur were found by B. Olsonowski and J. Wyatt Durham in San Luis Obispo County, having weathered out of a limestone lens in Late Jurassic rocks of the Franciscan Formation. The vertebrae were identified by Samuel Welles and given the name *Plesiosaurus hesterus* (Welles 1953). Welles later admitted that this designation was premature; because it is impossible to name a plesiosaur from one or two vertebrae, the specimens should simply be designated as from an unidentified plesiosaur (Welles n.d.). They appear to come from a creature that had the proportions of a large dolphin.

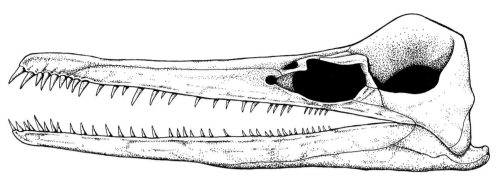

FIGURE 4.26
Dolichorhynchops skull. After Carpenter 1989.

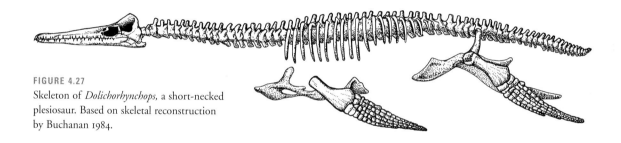

FIGURE 4.27
Skeleton of *Dolichorhynchops,* a short-necked plesiosaur. Based on skeletal reconstruction by Buchanan 1984.

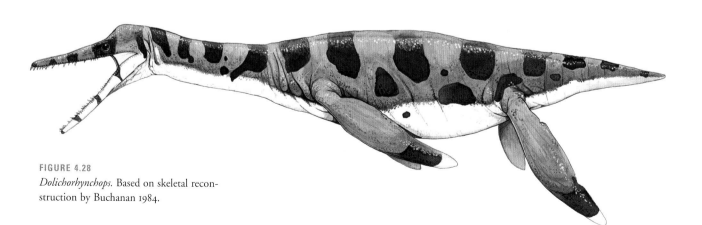

FIGURE 4.28
Dolichorhynchops. Based on skeletal reconstruction by Buchanan 1984.

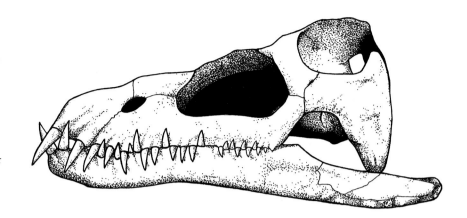

FIGURE 4.29
Skull of *Cryptoclidus*. After
Brown, in Norman 1985.

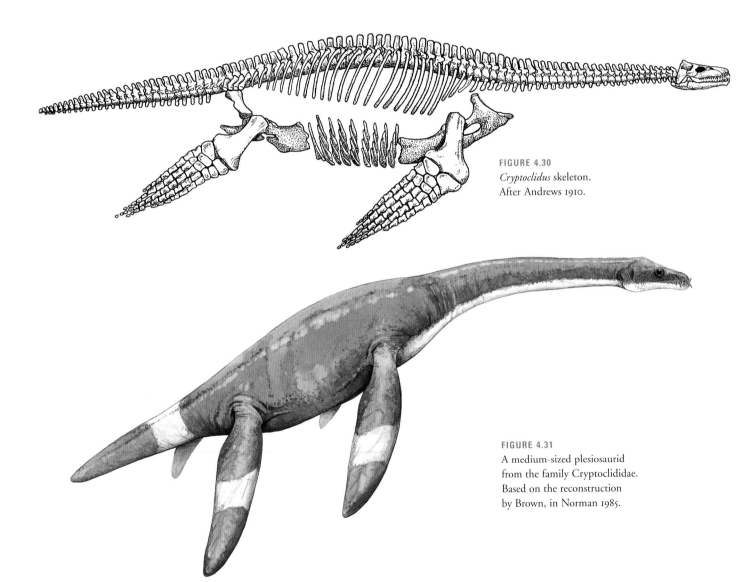

FIGURE 4.30
Cryptoclidus skeleton.
After Andrews 1910.

FIGURE 4.31
A medium-sized plesiosaurid
from the family Cryptoclididae.
Based on the reconstruction
by Brown, in Norman 1985.

Welles described this find as "the first Jurassic plesiosaur west of the Rocky Mountains." Certainly it is the first to be described, but in 1913, as noted above, Thomas Knock had found several vertebrae and ribs of what may have been a Late Jurassic plesiosaur (personal letters of John C. Merriam, Bancroft Library, UC Berkeley).

In 1997 Welles was working on a plesiosaur found in the Mariposa Formation in Mariposa County. He was writing a paper on this specimen just before his death. It is tentatively listed as being from the family Cryptoclididae. Plesiosaurs from this family have been found in rocks that range in age from Late Jurassic to Late Cretaceous (Brown 1981). Most species of this family are medium-sized plesiosaurs about ten to twelve feet in length.

Finally, fragmentary remains of an elasmosaur (a long-necked plesiosaur), consisting of a small string of vertebrae, have been found in the Bedford Canyon Formation of Orange County.

Cretaceous Plesiosaurs

Late Cretaceous plesiosaur remains exist from San Bernardino County in southern California and the hills along the western edge of the San Joaquin Valley. A single tooth, possibly from a plesiosaur, was found near Oroville in Butte County in northern California. From these fossils, four species have been identified, all long-necked plesiosaurs called elasmosaurs (Welles 1952).

Chester Stock (1942) of the California Institute of Technology in Pasadena stated that elasmosaurs were "sometimes likened to a turtle with a snake drawn through its body." However, they did not have so flat a body as a turtle, nor did they have a shell. They did have a long, snakelike neck with a relatively small skull mounted on a sturdy body, with large paddles and a short tail. The paddles were for swimming, and it is likely that the neck was used like a snake's body for quick "striking" at their swimming prey. Fish bones have been found in their digestive area, indicating that fish were at least part of their diet. Sometimes as much as "two handfuls" of polished gizzard stones (gastroliths) up to an inch in diameter have been found in the digestive cavity (Stock 1942). Williston (1902) reported finding over eight quarts of gastroliths in one Kansas specimen. Some dinosaurs employed gastroliths to grind up coarse food just as some modern birds swallow grit to grind hard seeds. However, the unpulverized remains of prey found with the gastroliths in some plesiosaurs (Brown 1904) have led some to suspect that they may have been used for ballast rather than digestion (Darby and Ojakangas 1980). Plesiosaurs may have swallowed just enough stones to become virtually weightless in the water, whereupon they proceeded to use their fins to "fly" wherever they pleased. Crocodiles swallow stones in this fashion to neutralize their buoyancy and to help stabilize their bodies in currents.

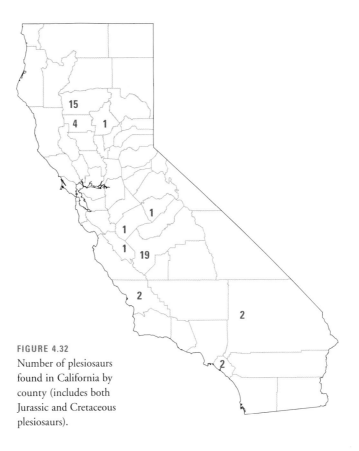

FIGURE 4.32
Number of plesiosaurs
found in California by
county (includes both
Jurassic and Cretaceous
plesiosaurs).

Aphrosaurus furlongi (LACM/CIT-2748) was found in the Late Cretaceous Moreno Formation in the Panoche Hills of Fresno County by Robert Leard. The specimen consisted of
several vertebrae along with the bones of the pelvic area and both rear paddles. The term
Aphrosaurus means "sea foam reptile." It was named *furlongi* for Eustace Furlong (John C.
Merriam's field assistant and preparator while he was at UC Berkeley, and preparator for
Chester Stock at Caltech). Welles (1943) described *Aphrosaurus furlongi* as a long-necked plesiosaur that was probably less active than *Morenosaurus.*

Also discovered by Robert Leard in the Late Cretaceous Moreno Formation of the Panoche
Hills of Fresno County was an incomplete skeleton (portions of the shoulders, hips, and
paddles) of *Fresnosaurus drescheri* (LACM/CIT-2758). *Fresnosaurus* means "reptile of Fresno
(County)," and the specific name *drescheri* was in honor of Arthur Drescher, who worked
as head of Chester Stock's field crews from 1939 to early 1941. Welles (1943) mentioned that,
although the actual neck length of *F. drescheri* is unknown, it is probably a long-necked plesiosaur. Welles (1952) said that the specimen is a juvenile but thinks there are enough defining
features to separate it from any other plesiosaur.

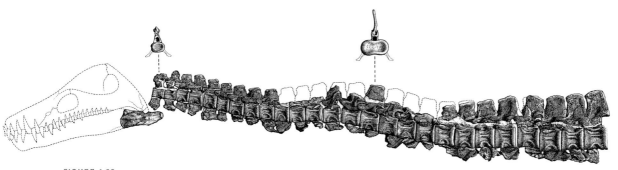

FIGURE 4.33
Aphrosaurus furlongi neck
vertebrae. From Welles 1943.

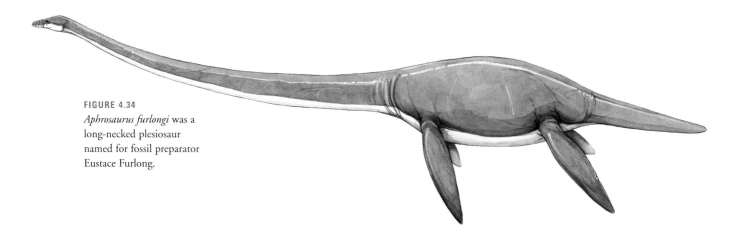

FIGURE 4.34
Aphrosaurus furlongi was a
long-necked plesiosaur
named for fossil preparator
Eustace Furlong.

Hydrotherosaurus alexandrae (UCMP-33912) was described by Welles in 1943. *Hydrothero* means "fisherman," while *saurus* refers to "reptile"; it was named for University of California Museum of Paleontology patron Annie Alexander. The specimen was discovered in the Panoche Hills in the Late Cretaceous Moreno Formation by rancher Frank C. Paiva and represents one of the most complete plesiosaur skeletons ever found. Welles (1943) describes this animal as an active, fish-eating plesiosaur with a long, flexible neck. It was over twenty-three feet long, about fifteen feet of which constituted the head and neck. The skull is one and a half feet in length, with long, sharp, curved teeth projecting from both its lower and upper jaws (Welles 1952). Casts of *H. alexandrae* are on display at McLane Hall at California State University, Fresno; the W. M. Keck Museum, located in the Mackay School of Mines Building at the University of Nevada, Reno; the California Academy of Sciences in San Francisco; and the new Life Sciences Building at the University of California, Berkeley.

FIGURE 4.35 (RIGHT)
Cast of *Hydrotherosaurus alexandrae* at the California Academy of Sciences, San Francisco. This specimen is one of the most complete plesiosaur skeletons ever found. Photo by the author.

FIGURE 4.36 (MIDDLE)
Hydrotherosaurus alexandrae skeleton. After skeletal reconstruction by Owen J. Poe, in Welles 1943; courtesy of Donald Dupras.

FIGURE 4.37 (BOTTOM)
Skull of *Hydrotherosaurus alexandrae*. After skull reconstruction by Otto, in Welles 1943.

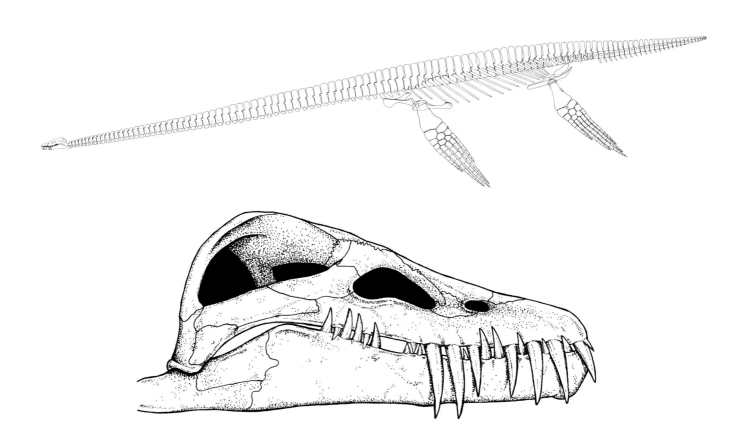

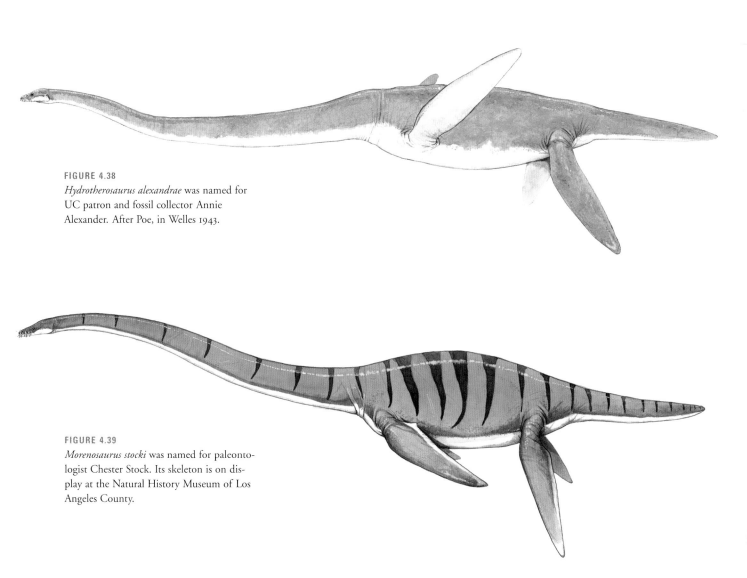

FIGURE 4.38

Hydrotherosaurus alexandrae was named for UC patron and fossil collector Annie Alexander. After Poe, in Welles 1943.

FIGURE 4.39

Morenosaurus stocki was named for paleontologist Chester Stock. Its skeleton is on display at the Natural History Museum of Los Angeles County.

Welles (1943) described *Morenosaurus stocki* (LACM/CIT-2802) as a very active, relatively short-necked plesiosaur. The skeletal remains were found by Robert Wallace and Arthur Drescher in the Panoche Hills of Fresno County in Late Cretaceous marine rocks of the Moreno Formation, for which it was named. The species name is for Chester Stock, who organized the field crews from the California Institute of Technology. It was described from a fairly complete skeleton that lacked only the head, part of the neck, and parts of its paddles (Welles 1952). Nestled in its digestive area were polished gastroliths. Welles (1952) estimated that the animal was over twenty-two feet long. Working under the direction of Stock, William Otto and Eustace Furlong made a full thirty-foot-long skeletal mount from these remains. It was put on display at Caltech in 1943; today one can see it at the Natural History Museum of Los Angeles County.

MOSASAURS

Mosasaurs were large (four to forty-four feet long) seagoing relatives of lizards whose skeletal remains have been found on all continents plus New Zealand and Timor (Camp 1942a; Bell 1993). In California their remains have been found in Butte, Placer, Fresno, Stanislaus, and San Diego Counties. The last appearance of mosasaur remains in the fossil record occurs in Missouri in what is interpreted to be tsunami debris caused by the impact event that ended the Mesozoic Era (Campbell and Lee 2001).

Mosasaurs lived only in the Late Cretaceous, making their appearance very late in the Mesozoic marine environment. They evolved originally from a subgroup of the varanoid lizards that included the closely related marine lizards called the aigialosaurs. It has been suggested by Russell (1967) that their late evolution must have been a result of new niches opening up in the near-surface carnivorous environment, perhaps as a result of the disappearance of ichthyosaurs. The still-abundant plesiosaurs fed on fish close to the shallow bottom, while most mosasaurs were efficient surface feeders and usually not in direct competition with plesiosaurs (Russell 1967). Recent evidence, however, suggests that some mosasaurs may have frequented deep water (Sheldon 1997). In addition, Gorden Bell (1996) has found evidence that mosasaurs may have fought each other as rivals, and some may have chosen mosasaur species other than their own for an occasional meal.

Mosasaurs had large heads and short necks mounted on long flexible bodies. They had eel-like tails and well-developed flippers, perfect for agile swimming. Mosasaurs swam like marine iguanas, using undulations of the back portion of their bodies and tail. The flippers were primarily for steering and quick turns. Their relatively large size and good maneuverability probably enabled mosasaurs to overtake and seize smaller, streamlined prey. Mosasaur mouths were studded with sharp conical teeth having keeled edges, and some were even serrated (Bell 1996). Like monitor lizards and snakes, the lower jaw had a curious second hinge about three-quarters of the way back. This allowed these animals to open their mouths wide, enabling them to catch, dismember, and swallow their prey more effectively. They probably had well-developed eyesight and excellent hearing. All these evolutionary advancements were perfect for consuming fish and swimming mollusks such as ammonites and squid. Differences in size as well as body and fin shape indicate that each species of mosasaur evolved differently to exploit various ways of making a living (Russell 1967).

The recent discovery of two apparently prenatal, very small mosasaurs associated with an adult of the same species suggests that at least some species bore live young instead of having to return to the shore to lay eggs (Bell and Sheldon 1996). Histological evidence, along with the large embryonic size and the lack of any eggshell material, supports live birth as its

FIGURE 4.40
Mosasaurs found in
California (number
by county).

reproductive mode (Sheldon 1996). Gorden Bell (pers. comm. 2000) says they have now found evidence of two more apparently unborn fetuses from that same adult mosasaur.

Two of the California mosasaurs have been placed in the genus *Plotosaurus*. With their narrower flippers and more highly evolved caudal (tail) fins, plotosaurs were probably faster swimmers than other mosasaurs (Russell 1967). These plotosaurs described by Camp (1942a) are geologically the youngest articulated mosasaur skeletons found anywhere, and based on cladistic analysis *Plotosaurus* is the most distinctly highly evolved mosasaurid found thus far (Bell 1997).

Plotosaurus bennisoni (UCMP-32778) was originally named *Kolposaurus bennisoni* by Camp (1942a), but the genus name had already been used on another animal (Camp 1951). The original name, *Kolposaurus,* means "estuary (or bay) reptile," while *Plotosaurus* means "swimming reptile." *Plotosaurus bennisoni* was named for Allan Bennison, the boy who found it in 1937 in the Late Cretaceous beds of the Moreno Formation in the hills along the western side of the San Joaquin Valley in Merced County. The specimen included several vertebrae,

FIGURE 4.41 (TOP)
Skull of *Plotosaurus bennisoni.* Courtesy of the University of California Museum of Paleontology; photo by the author.

FIGURE 4.42 (MIDDLE)
Skeleton of *Plotosaurus bennisoni.* From Camp 1942a; courtesy of Donald Dupras.

FIGURE 4.43 (BOTTOM)
Plotosaurus bennisoni is named for its discoverer, Allan Bennison. After skeletal reconstruction by Poe, in Camp 1942a.

some ribs, and the finest mosasaur skull ever found in California. The relatively large eyes were set far back along the sides of its sixteen-inch skull, and the snout bore jaws filled with sharp, meshing teeth (Camp 1951).

Another plotosaur, *Plotosaurus* (originally *Kolposaurus*) *tuckeri,* was found in the Panoche Hills in 1937 by rancher Frank Paiva and Professor William M. Tucker of Fresno State College. It was named for Tucker, who had also been instrumental in helping bring *Hydrotherosaurus alexandrae* to light. *Plotosaurus tuckeri* is represented in the fossil record by a fairly

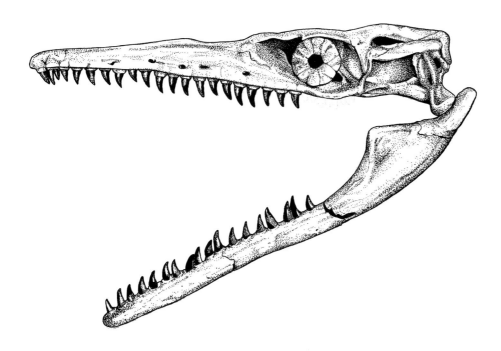

FIGURE 4.44 (LEFT)
Skeleton of *Plotosaurus tuckeri*. Courtesy of the Natural History Museum of Los Angeles County; photo by the author.

FIGURE 4.45 (BELOW)
Skull of *Plotosaurus tuckeri*.

large skull and the front half of the skeleton, including the shoulder girdles and paddles. Fish remains were found in the abdominal area.

Upon detailed analysis of the 26.5-inch skull, Camp (1942a) noted that it was not as advanced toward aquatic adaptations as *Plotosaurus bennisoni*. In 1993 Bell noted the close similarity between *P. tuckeri* and *P. bennisoni,* but in 1997 he showed them to be individual species

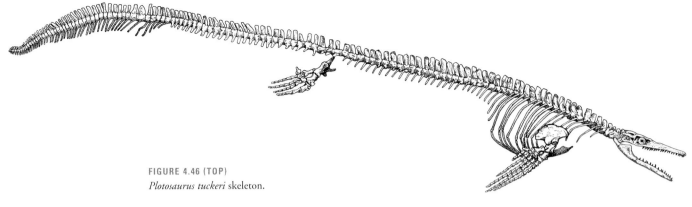

FIGURE 4.46 (TOP)
Plotosaurus tuckeri skeleton.

FIGURE 4.47 (BOTTOM)
Plotosaurus tuckeri is named
for Fresno State College
geologist William Tucker.

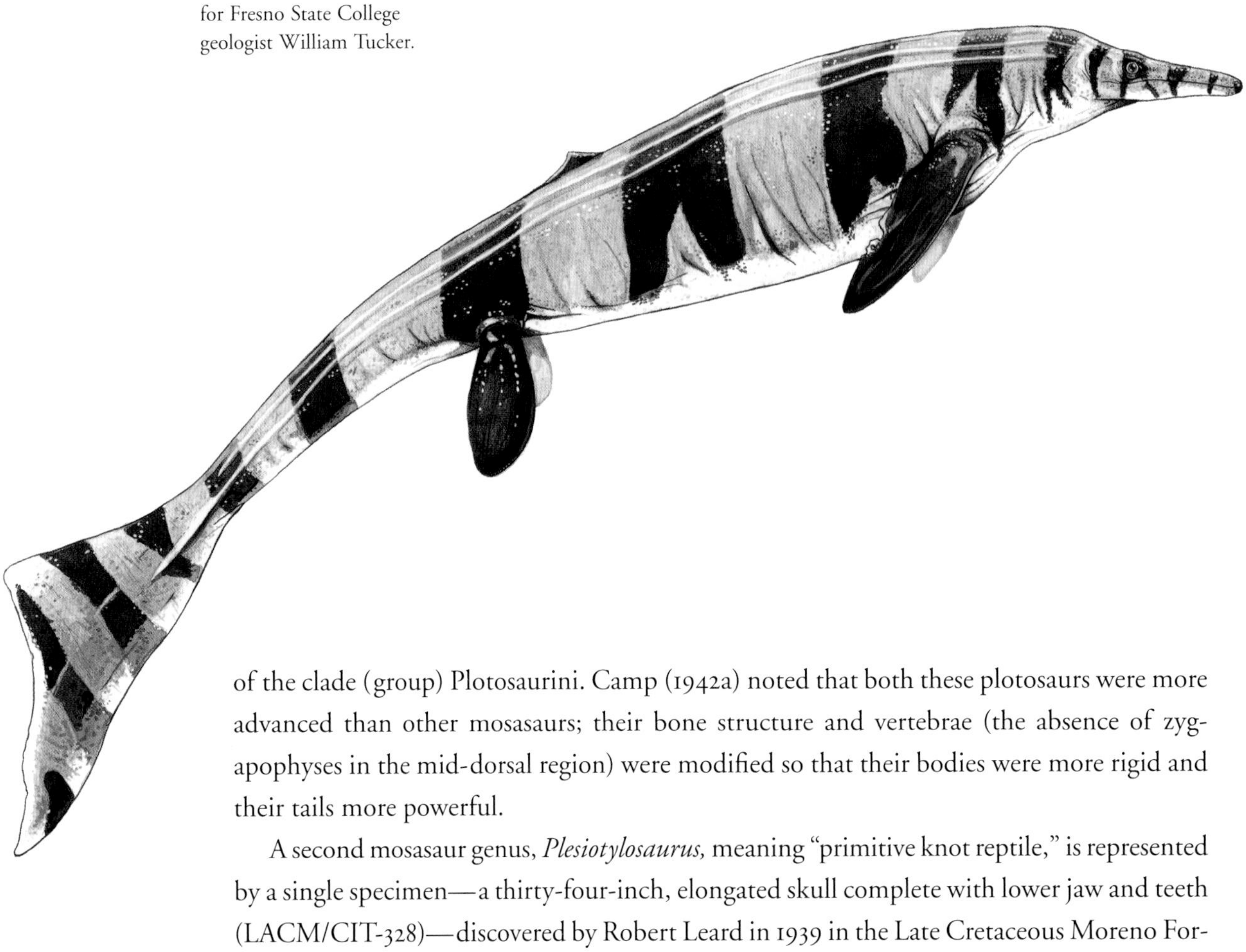

of the clade (group) Plotosaurini. Camp (1942a) noted that both these plotosaurs were more advanced than other mosasaurs; their bone structure and vertebrae (the absence of zyg-apophyses in the mid-dorsal region) were modified so that their bodies were more rigid and their tails more powerful.

A second mosasaur genus, *Plesiotylosaurus,* meaning "primitive knot reptile," is represented by a single specimen—a thirty-four-inch, elongated skull complete with lower jaw and teeth (LACM/CIT-328)—discovered by Robert Leard in 1939 in the Late Cretaceous Moreno For-mation in the Panoche Hills of Fresno County. Christening it *Plesiotylosaurus crassidens,* Camp (1942a) described this huge mosasaur as being like the mosasaur *Tylosaurus* but with larger

FIGURE 4.48 (LEFT)
Thirty-four-inch skull of
Plesiotylosaurus crassidens.

FIGURE 4.49 (BELOW)
Plesiotylosaurus crassidens was
a huge mosasaur.

pterygoid teeth, shorter and stouter quadrates, expanded prefrontals, and without the extended rostrum.

In January 1994 Patrick Antuzzi found portions of a skull (SC-VR59) of a moderate-sized mosasaur in the Chico Formation of Placer County (Hilton and Antuzzi 1997). The complete skull would have been at least twenty-five inches long. Its sharp, keeled teeth were slightly backward-curving. According to paleontologist Gorden Bell of the National Park Service, evidence from the quadrates (the bone that connects the jaw to the skull) suggests that the specimen is different from others found in North America. Possibly a derived species of *Clidastes* not previously recognized in North America, it may be closely related to a form thus far found only in New Zealand (Bell, pers. comm. 2000). The proportional distance between the front and back flippers of *Clidastes* is longer than on any other mosasaur.

According to Dianne Yang (1983), careful examination and comparison of mosasaur pectoral and pelvic appendages from the Panoche Hills indicates that other, undescribed mosasaurs may remain from the region. Perhaps renewed work in the area will bear this out.

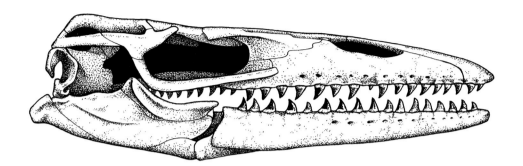

FIGURE 4.50
Skull of *Clidastes*. After
Russell 1967.

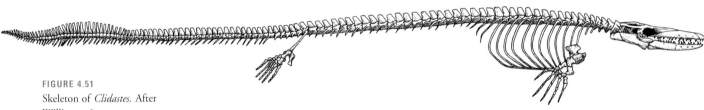

FIGURE 4.51
Skeleton of *Clidastes*. After
Williston 1893.

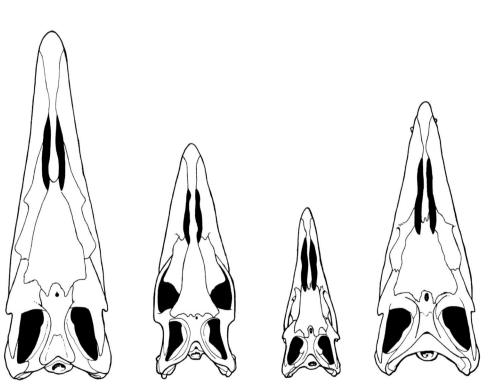

FIGURE 4.52
Skulls (from left to right)
of *Plesiotylosaurus crassidens,
Clidastes, Plotosaurus bennisoni,* and *Plotosaurus tuckeri*. From Camp 1942a and
Williston 1893.

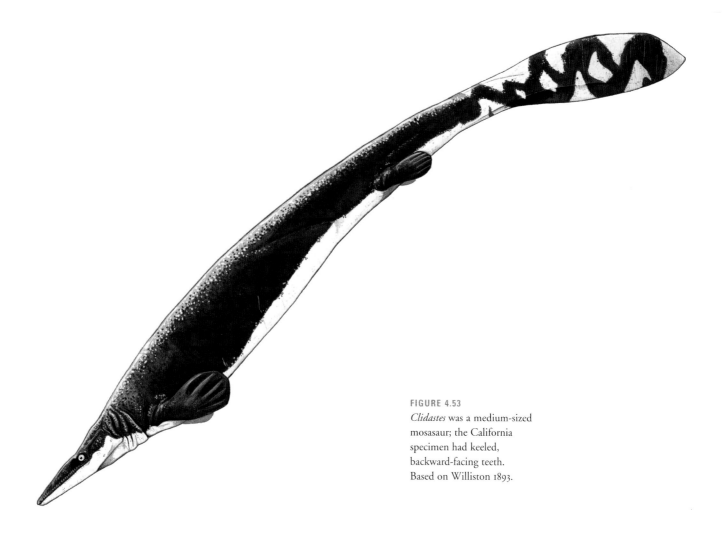

FIGURE 4.53
Clidastes was a medium-sized
mosasaur; the California
specimen had keeled,
backward-facing teeth.
Based on Williston 1893.

TURTLES

Turtles have traditionally been grouped with anapsids because, unlike most reptiles, they lack holes in their skulls in back of the eyes; as a consequence, they were thought to come from more primitive reptilian stock. This idea is now being challenged, however, and turtles may be more advanced than once thought. It is conceivable they are more closely related to the diapsids, which have two holes in their skulls, and perhaps even to the Sauropterygia, the group to which the very sophisticated reptiles called plesiosaurs belong (Rieppel 1999).

Turtles are known in rocks since the Late Triassic, and several Mesozoic specimens have been found in California, all from Late Cretaceous marine rocks. Fossil turtles have been found in Butte, Orange, San Diego, Fresno, Placer, and Tehama Counties. Most are known only from fragmentary remains, and thus far none have been identified any further than to genus.

FIGURE 4.54
Fossil turtles found in
California (number by
county).

Nearly all of the Cretaceous turtles found in California were marine; however, one of them, *Basilemys,* may actually be a land-based turtle, and another, *Naomichelys,* may have frequented freshwater environments. Both may have simply washed out to sea in death, then sunk to the bottom and become fossilized.

A turtle first described from a skull, mandible, limb bones, and carapace was found by Chad Staebler in 1979 in the Late Cretaceous Moreno Formation in the Panoche Hills of Fresno County. At the time it was identified to the genus *Adocus* by J. Howard Hutchison of the University of California Museum of Paleontology. Recently Hutchison placed this specimen (UCMP-126717), as well as a new find (UCMP/ETC-GC958122) from Orange County, within the genus *Basilemys,* in the family Nanhsiungchelyidae (Hutchison, pers. comm. 2000). The Orange County specimen was found by Gino Calvano in the Late Cretaceous Ladd Formation during excavation of the Eastern Transportation Corridor, a pay-as-you-go "freeway" constructed against the foothills of the Santa Ana Mountains (Roeder, pers. comm. 1997; Lander 2002).

The family Nanhsiungchelyidae appears first in the fossil record in the Early Cretaceous of Asia (Nessov 1984) and later spread into North America in the Late Cretaceous. The North American nanhsiungchelyids are all referred to the genus *Basilemys* (see figs. 4.55 and 4.56). Both California specimens of *Basilemys* were found in Late Cretaceous marine rocks, and Hutchison (pers. comm. 2000) thinks that these were terrestrial individuals that probably accidentally got swept down rivers or streams into the ocean. These two specimens, the first to be found on the Pacific slope, are different enough to be separable from all others of their kind (Hutchison, pers. comm. 2002). Other scientists agree that members of the genus *Basilemys* were terrestrial herbivorous turtles (Brinkman 1998; Mlynarski 1972). According to Hutchison, they were very much like living tortoises, with elephantine feet and a blunt tortoise-like face. Others argue, however, that *Basilemys* may have been a group of aquatic turtles that were specialized swimmers, using their forelimbs to move and cling to the bottom in strong currents (Sukhanov and Narmandakh 1977). Hutchison thinks that the features these investigators emphasize are probably primitive ones retained from aquatic ancestors (pers. comm. 2002). He points out that while tortoises (such as those of the Galápagos) may sit around in ponds of water to keep cool or may even live in swampy areas, they can hardly be regarded as aquatic for this reason. In *Basilemys,* the graviportal features (weight-bearing characteristics) of the limbs, gross morphology of the shell, extensive skin armor, and development of a gular projection of the plastron (a projection below the head and neck on the lower shell) argue most strongly for tortoiselike adaptation (Hutchison, pers. comm. 2002).

Among the smaller reptilian material found in Baja California, Mexico, and sent to Richard Estes of San Diego State University for study (Morris 1974b) were bones from "a pustulse turtle." Hutchison reports (pers. comm. 2000) that this is in the genus *Naomichelys,* a type of turtle usually found from brackish to marine environments or in freshwater environments near the coastline. According to Hutchison, this turtle, which is known in the published literature only from shell fragments, appears to be related to the European *Helochelys.*

The remainder of the turtle fossils from California are from the superfamily Chelonioididea, which encompasses all of the truly marine turtles. These turtles use their limbs (which have been modified into paddles) to "fly" through the water. Chelonioids, which have been found in rocks from the Early Cretaceous to the present, were much more morphologically diverse in the Cretaceous than today, a reflection, perhaps, of greater ecological diversity in Cretaceous times (Hirayama 1997). Three families of chelonioid turtles have been found in California: Dermochelyidae, Cheloniidae, and Toxochelyidae.

The dermochelyid turtles include the modern leatherback, the only living sea turtle that is truly an open-ocean turtle (see fig. 4.57). As the name implies, instead of having a hard carapace leatherbacks have a leathery skin stretched over broad riblike structures running

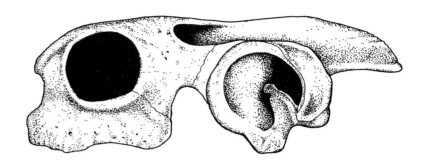

FIGURE 4.55 (RIGHT)
Basilemys skull. Based
on reconstruction by
Brinkman 1998.

FIGURE 4.56 (BELOW)
Basilemys may have been
a land-based turtle.
Based on reconstruction
by Brinkman 1998 and
Langston 1956.

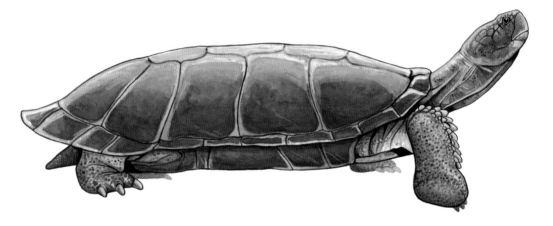

lengthwise along the "shell." In other turtles and their land-dwelling cousins, the tortoises, the shell is formed largely by a fusion of dermal bone and ribs, but in modern leatherbacks this is not the case. The original ancestral shell was formed much like that of any normal turtle, but this dwindled through time. When a new "shell" evolved it was not from the ribs but from an array of tilelike dermal (skin) plates (much like those of the giant paleo-armadillo called the glyptodont) over which the leathery skin stretched. The leatherback has been known to dive to depths of about three thousand feet (McAuliffe 1995).

The first evidence of a dermochelyid turtle (see fig. 4.58) to be found in California was a coracoid (shoulder bone) (SC-VRT19) found by Patrick Antuzzi in the Late Cretaceous marine layers of the Chico Formation in Placer County (Hilton and Antuzzi 1997). The size of the bone indicates a turtle that would have had a carapace about four and a half feet in length. A scapula, or shoulder blade (UCMP-172070), belonging to a smaller animal was found in the Chico Formation in Butte County by Eric Göhre.

FIGURE 4.57 (LEFT)
Leatherback sea turtle in 1923 on the dock in Monterey, California. Photo from the collection of the author.

FIGURE 4.58 (BELOW)
Dermochelyid turtles were marine turtles. Their only living relative is the leatherback turtle.

Both of these specimens appear to be from turtles more closely related to modern leatherback turtles than to the Cretaceous *Mesodermochelys.* It is likely that these two bones are from a new, as yet undescribed species. They are, moreover, the oldest known dermochelyids from the Eastern Pacific region (Parham and Stidham 1999).

The family Cheloniidae is the largest group of chelonioids in the fossil record. Except for

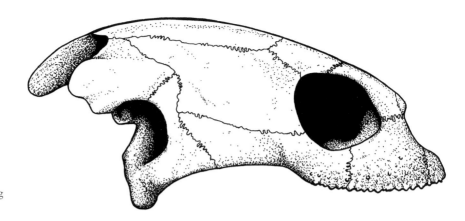

FIGURE 4.59 (RIGHT)
Osteopygis skull. After
Foster 1980.

FIGURE 4.60 (BELOW)
Osteopygis had an abnormally large skull with a
broadly rounded snout,
perhaps used for crushing
ammonites.

the leatherback turtle, most living sea turtles are from this family (Hirayama 1997). The carapace of most members of the family is incomplete compared to more typical turtles, in that
the fusion of dermal bone and ribs does not completely enclose the animal. Toward the skirt-
ing around the edge of the shell these fused bones narrow down, leaving open spaces in the
bone structure (see fig. 4.63) (Zangerl 1953).

A portion of a skull (supraoccipital) (SC-VRT-32) from a turtle in this family was found
in the Late Cretaceous Chico Formation in Granite Bay by Patrick Antuzzi. This is the oldest Pacific occurrence of this family, and although Cheloniidae and Dermochelyidae are found

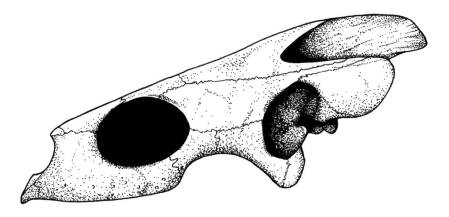

FIGURE 4.61
Skull of a toxochelyid turtle. After Zangerl 1953.

in association with each other in Cenozoic sediments, this is the first association of these two families in the Mesozoic fossil record (Parham and Stidham 1999).

In 1979 Chad Staebler discovered a marine turtle from the genus Osteopygis (UCMP-123616) in the Late Cretaceous Moreno Formation in the Panoche Hills of Fresno County. This was the first osteopygine to be found on the west coast of North America. The specimen consisted of a complete skull and jaw, two humeri (upper flipper bones), a scapula (shoulder blade), three peripheral plates (portions of the carapace), plus other bits and pieces (Foster 1980). The shell of *Osteopygis* (see fig. 4.65) consists of thick plates and is rather oval and moderately arched (Zangerl 1953). This animal has an abnormally large skull with a broadly rounded snout, which is thought to have been used to crush hard-shelled prey such as ammonites (Parham, pers. comm., 1999; Zangerl 1953).

The family Toxochelyidae is represented in California by a single specimen, a skull fragment (supraoccipital) (SC-VRT32) found by the author in the Late Cretaceous Chico Formation in Granite Bay and identified as a "toxochelyid"-grade cheloniid (Parham and Stidham 1999). Toxochelyids are primitive, extinct marine turtles found from Late Cretaceous rocks of North America. They were specialized for bottom feeding in shallow-water, nearshore environments (Hirayama 1997). Toxochelyids were possibly the most common type of sea turtle in Late Cretaceous seas east of the Rocky Mountains.

The shell of toxochelyids is relatively light in construction and is commonly rather circular in outline, though some may be more oval. The skull is roughly triangular, and the

FIGURE 4.62 (TOP)
Toxochelyids were primitive
marine turtles specialized for
bottom feeding in shallow
nearshore environments.
Partially modified from skull
illustration in Zangerl 1953.

FIGURE 4.63 (BOTTOM)
Comparison of the cara-
paces of (a) *Osteopygis,* (b)
Basilemys, (c) dermochelyid,
and (d) toxochelyid turtles.
After Hirayama 1997;
Langston 1956; and
Zangerl 1953.

A

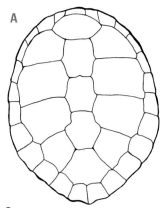

B

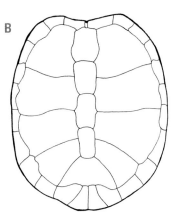

C

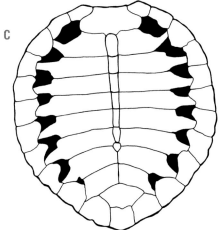

D

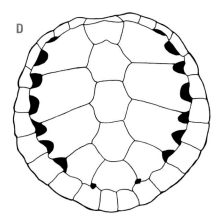

bones of the skull roof are arranged much as in recent marine turtles. It has a rather blunt snout. Its front limbs are modified as primitive flippers. Locomotion in these animals appears to be transitional between that of common pond turtles and sea turtles, suggesting that toxochelyids were capable of short bursts of speed and were fairly efficient at cruising through the water. As in today's sea turtles, cruising was most likely powered by its longer flipperlike front limbs, while bursts of speed were probably accomplished by rapid paddling of the hind limbs, as in pond turtles (Zangerl 1953).

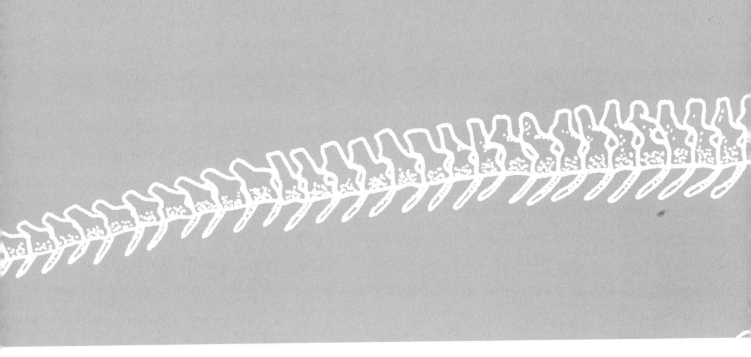

PART II

THE HISTORY OF DISCOVERY

THE DISCOVERERS of Mesozoic reptile remains are for the most part scientists, although a few are amateurs. (I use the word *scientist* rather than *paleontologist* because many of the people involved in fossil hunting were not formally schooled in paleontology.) Scientists, with their knowledge and training, study fossil remains and then publish information about them. This brings the discovery to the scientific community, where other scientists might further study the specimen and perhaps gain even more information from it and make new interpretations. But scientists and amateur bone hunters are only part of the story. From the finding of a new fossil to its description and illustration in a scientific journal takes the work of several

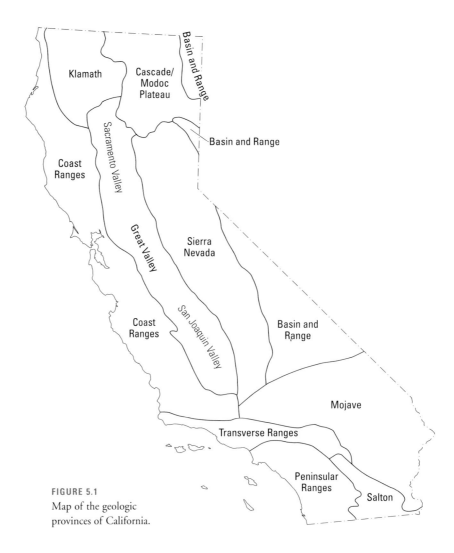

FIGURE 5.1
Map of the geologic
provinces of California.

types of people, and in the following two chapters I will try to highlight their participation as much as possible.

Just as important as the discoverer is the preparator, whose job is to delicately work the fossil from its rocky matrix for study and possible display. This can be a tedious affair requiring much skill and patience as rock is carefully stripped away from the specimen. These hardworking and highly talented people often receive too little credit, with most of the glory going instead to the scientist who studies and may publish information on the specimen.

In addition, an artist has the task of taking the scientific information about the fossil and giving it life in an illustration or sculpture. An accurate rendering of what the animal may have looked like makes it more real to all of us. Another important person is the photographer, who aids in documenting the processes of excavation and preparation. Often the paleontologist serves as the photographer, knowing the proper lighting and shadow needed to accentuate anatomical details.

The chapters that follow, a historical journey featuring the people who made the discoveries and who brought the fossils to light, are organized geographically, using the system of geologic provinces. These provinces are based mostly on rock types and ages plus geologic structure. Fossil Mesozoic reptiles have been found in all of California's geologic provinces with the exception of the Basin and Range, Cascade/Modoc, and Salton Provinces.

Most of the rocks in the Salton Province (named for the Salton Sea, and sometimes called the Salton Trough) were laid down in the last 20 million years, long after the Mesozoic. The Cascade/Modoc Province is underlain by rocks of the Mesozoic Great Valley Group and by rocks typical of the northern Sierra and eastern Klamath that could yield Mesozoic reptile remains, though extensive coverage by younger lavas makes such discoveries unlikely. (I did find one turtle fossil in Great Valley Group rocks beneath Cascade lavas, but I have included it in the Great Valley Province.) The Basin and Range Province has yielded Mesozoic reptile remains in Nevada but not yet in California.

The Peninsular Ranges, Transverse Ranges, Mojave Desert, Coast Ranges, Great Valley, Sierra Nevada, and Klamath Mountain Provinces have all yielded important Mesozoic reptile remains. The portion of the Peninsular Range that extends into Baja California has provided many more, giving us a rich heritage in Mesozoic reptile fossils in California. We start in the Klamath Mountains at the top of the state, where the first discoveries were made, and work south, finishing in the province of Baja California Norte, in Mexico.

5

THE NORTHERN PROVINCES

THE NORTHERN PROVINCES encompass all of California north of the Transverse Ranges, which dissect the state in an east-west fashion just north of the Los Angeles Basin. We begin our historical tour of discoveries in the geological northern extension of the Sierra Nevada called the Klamath Mountains Province. It is in this province that all the numerous Triassic marine reptile finds have been made and much of the stimulating early history of discovery happened. We then slip south to the Sierra Nevada, where, despite the scarcity of Mesozoic reptiles, the perseverance of scientists has paid off. From the Sierra we go westward to the long Great Valley Province, where a multitude of discoveries has yielded a rich trove of Late Jurassic and Cretaceous dinosaurs and marine reptiles. Finally, we visit the jumble of the Coast Ranges Province; although the finds here have been few and far between, they provide us with interesting accounts and intriguing fossils.

KLAMATH MOUNTAINS PROVINCE

All of the more than two hundred Triassic reptile fossils found in California have come from the Klamath Mountains Province and its Upper Triassic (ca. 205 MYBP) Hosselkus Limestone. The Klamath Mountains Province in California lies south of the Oregon border, west of the Cascade chain of volcanoes, north of the Sacramento Valley, and east of the Coast Range Thrust, a fault just inland from the ocean. It is the northern extension of the older rocks of the original Sierra Nevada, but unlike most of the rocks of the Sierra, these rocks have not been as badly heated and pressed. As a result, more fossils have survived. Between the Sierra and Klamath Provinces, the Sierran rocks are buried by the younger, overlying, volcanic rocks of the Cascade Range Province. Looking in the deep

FIGURE 5.2

A Late Triassic reef scene, where basking thalattosaurs have used their
claws to climb out on the reef while an ichthyosaur feeds below.

stream-cut canyons of this portion of the Cascades one can often find windows cut into the older Sierran rocks below.

Much of the Klamath Mountains Province is very remote and rugged, covered with dense vegetation. To get to the most productive fossil sites today means a perilous drive of several hours in a four-wheel-drive vehicle on narrow, steep roads overgrown with scratching trees and brush. The summers, when fieldwork is typically conducted, are hot and dry. Poison oak abounds, and there are ticks, rattlesnakes, skunks, and even occasional mountain lions and bears. It was even wilder a hundred years ago.

Merriam and Cohorts: 1901 to 1910

From 1901 to 1910 John C. Merriam of the University of California, Berkeley, launched ten or eleven expeditions to locate Mesozoic reptiles in the Hosselkus Limestone. Merriam's team also successfully searched for fossil Pleistocene (Ice Age) mammals in caves in Shasta County during this same period. These expeditions were largely funded by Miss Annie Alexander, heir to a massive fortune accumulated through sugar plantations in Hawaii and the Matson Shipping Lines. In 1900 Annie attended Merriam's lectures on paleontology and became fascinated with his discoveries. On the condition that she could go along, she offered financial support for his summer fieldwork.

The expeditions to Shasta County came about as the result of the work of Professor James Perrin Smith of Stanford University. In 1893, while working for the U.S. Geological Survey collecting ammonites, he stumbled on Triassic reptilian remains in the Hosselkus Limestone. Smith collected several vertebrae, some ribs, and a couple of limb bones. In 1894 he published the first scientific article to mention Mesozoic reptile fossils in California. He referred to the bones he had discovered as being from a *Nothosaurus,* a relative of plesiosaurs known mainly from Triassic rocks of Europe. Smith gave the bones to Merriam for study in the spring of 1895, and after careful scrutiny Merriam concluded that they belonged to an ichthyosaur about the size of a dolphin. He named the new creature *Shastasaurus pacificus.* At about the same time, an O. B. Sherman discovered reptilian remains in another area of the Hosselkus Limestone and brought them to Smith's attention.

In the summer of 1901 Smith took Merriam and a party of students to the fossil-rich outcrops, where they found more bones. That fall Alexander sponsored a two-month expedition led by UC Berkeley vertebrate fossil preparator Herbert W. Furlong. (Herbert later moved on to the Pacific Commercial Museum, in San Francisco [letter from Merriam, May 9, 1906], and was replaced by his younger brother, Eustace, who became an important fossil reptile discoverer and preparator for Berkeley and later for the California Institute of Technology;

JAMES PERRIN SMITH (1864–1931)

Although James Perrin Smith was not a vertebrate paleontologist and did not contribute much to the study of Mesozoic reptiles in California directly, he did discover and publish the first such specimen in the state. His find was thus the catalyst for discovery of all the Triassic finds in California.

Growing up in South Carolina just after the Civil War and steeped in the poverty that was common during Reconstruction, he was largely home-schooled. His family moved so that he could attend Wofford College in South Carolina, where he received his A.B. degree in 1884. He then went on to Vanderbilt University, where in 1887 he received his M.A. He taught high school math and science for two years, becoming interested in geology after reading Hugh Miller's *How to Know Rocks.*

After studying the subject on his own for a few years he was able to procure an appointment as assistant chemist and geologist on the new Arkansas Geological Survey under John Casper Branner. With Branner's encouragement Smith went to Göttingen, Germany, to study with Adolph von Koenen and Professor Liebisch. There he became interested in Mesozoic paleontology, especially Triassic ammonites, which he was finding in the strata of southern Bavaria. In 1892 he received his Ph.D., magna cum laude.

FIGURE 5.3

James Perrin Smith in 1930. Charcoal on paper by artist Peter Van Valkenburgh. From Plummer 1931.

FIGURE 5.4

Scapula (shoulder blade) labeled *Shastasaurus pacificus* previously housed at Stanford University. Courtesy of the California Academy of Sciences, San Francisco; photo by the author.

he is today known for his major contributions to the science of Mesozoic reptilian fossils as well as Pleistocene mammalian paleontology in California.) The group set up a base camp at the Matteson Ranch near the fossil-bearing limestones. For six days they hiked the two and a half miles of rugged terrain to and from the fossil site and lugged the fossils back. Tiring of this grueling work, they then procured horses to make the trip easier. They retrieved

Smith returned to the United States as an assistant professor of paleontology and mineralogy at the newly formed Stanford University—a choice influenced by the fact that California and Nevada possessed ample Triassic ammonite-bearing strata. In 1893 Smith began nearly thirty years of summer employment with the U.S. Geological Survey; this tenure led to his publication of the first comprehensive geologic map of California in 1894. During his career he became the expert on Triassic ammonites of the West Coast, and in searching the ammonite-rich Hosselkus Limestone of Shasta County he discovered the first Mesozoic reptile remains in the state. He informed vertebrate paleontologist John C. Merriam at the University of California of his discovery and in 1901 took Merriam to the outcrop. Smith's initial discovery inspired Merriam, Annie Alexander, Eustace Furlong, and others to seek further reptilian fossils. With over two hundred discoveries catalogued thus far, this region has relinquished more fossil reptiles than any other part of the state, and they are the only Triassic remains discovered in California.

Smith, according to a colleague, "was first and always a teacher . . . [and] was held in high regard by all of his students both in and out of the classroom" (Plummer 1931). Smith's good friend Solon Shedd described him as "one of the most kind and lovable men it has ever been [my] privilege to know. He was uniformly courteous, exceedingly modest and unassuming, and possessed the highest sense of honor" (Plummer 1931). It seems that the inspiration for Mesozoic reptilian discovery in California started with the right man.

parts of five skeletons and several scattered bones, mostly from Merriam's newly named genus of ichthyosaur, *Shastasaurus.*

In 1902 another expedition was launched. About this extremely productive expedition we have much more detailed information, which gives the full flavor of a collecting trip in early-twentieth-century California. The participants included Merriam, UC Berkeley assistant professor Vance C. Osmont, Merriam's preparator Eustace Furlong, and a Mr. Schaller, plus Alexander and her companion, Miss Katherine Jones. Merriam invited Miss Jones because it was considered inappropriate for Alexander, as a young single woman, to go into the field alone with a group of men. Jones was also single, about forty years old, and had taken all the paleontology courses at Berkeley (Stein 2001).

Miss Jones found some fossils, but more important for our purposes, she kept a detailed diary that allows us a colorful glimpse into the expedition. Their adventure began the evening of Sunday, June 15, 1902, when Jones, Alexander, and Osmont boarded the train at Oakland bound for Redding. The ferryboat at Port Costa that was to take the train across the Carquinez Strait was out for repairs, so they detoured to Stockton. This made them three hours late arriving in Redding; there they missed the stagecoach, so Miss Alexander had to hire a "rig" (horse and buggy) to take them to Winthrop. During this "glorious ride" they stopped to gather wildflowers. At Winthrop they met a Mr. Webb, who was collecting fish for Princeton and who had found the bone of a reptile. (It is unclear whether he was collecting native fish or fossil fish and

FIGURE 5.5
Partial skull of *Shastasaurus
alexandrae* (now *pacificus*).
Courtesy of the University
of California Museum of
Paleontology; photo by
the author.

whether the reptile bone was a fossil or not. Osmont purportedly got all the details on this possible fossil bone, but I find no trace of this early find in any of the collections.)

The next day after lunch at the hotel in Winthrop they rode on to the Matteson Ranch, where they procured horses to take them to a heavily wooded site near several springs and the fossil-bearing limestone cliffs. The following morning (June 17), to avoid the heat, they were on the limestone outcrops by 6 A.M. "It was pleasant in the morning," Jones noted, "but in the afternoon the sun beat mercilessly on our poor heads." The first site they looked at was where James Perrin Smith had collected ammonites and where a slab of rock containing "saurian" bones was lying fully exposed on the side of a hill. Taking shade under a pine tree they lunched on beans, crackers, and peaches and sipped much-needed water from their canteens. On their way back to camp that evening they passed the site where Osmont had discovered bone fragments at the base of the cliff, which he then traced to the two intact specimens of *Shastasaurus osmonti* above.

On the morning of June 18 they were off to work the cliffs in back of "the cove of ammonites." Here Alexander found a vertebra, while Jones discovered a tooth and something that looked like a rodent's jaw. Three more vertebrae and a rib were found that day, too. The next day the party separated, with the men working the cliffs and the ladies in the cove. They were in sight of each other, and in the clear stillness they could hear each other's voices. When something spooked a deer, Miss Jones commented how it covered as much ground in two minutes as it would take them twenty minutes to hike. Jones found two large pieces of lime-

FIGURE 5.6
Outcrops of gray Hosselkus
Limestone in the rugged
Klamath Mountains
Province. Photo by the
author.

stone full of bones, while the men above found six vertebrae, a rib, a limb bone, and another boulder full of bones.

They were back in the heat again on the twentieth, working hard on the slopes. Schaller found several vertebrae, Osmont a rib, and Alexander a rock with bone sticking out of it. Furlong split the rock and exposed a fine reptilian paddle (flipper) inside. The next day Alexander found a slab of rock containing ten vertebrae and other bones, while Furlong discovered limb bones and a slab with several other bones.

After a week of collecting they needed to pack some of the fossils out, so on Sunday the twenty-second they all headed for the Matteson Ranch. They loaded one of the horses with fossils, while Miss Jones rode the other; Annie walked the whole way and got sore feet as a result. On the trail they had to ford a creek, after which they rested under a cherry tree. Some of the party delighted in eating ripe cherries; meanwhile, Osmont caught five large trout and Annie sneaked downstream for a quick bath. She finished her bath none too soon, for while putting on her clothes a hunter happened on the scene. The next day found Alexander up at 4:30 A.M.; after the others woke and had breakfast they were off to hunt more fossils. They worked the hill by the cove in the morning but found nothing. In the afternoon they worked hard at gathering the remaining bones of *Shastasaurus osmonti,* though there was some doubt whether they were well enough preserved to take out.

On June 25 Osmont went to Matteson's ranch for supplies while the rest of the company looked for fossils along the trail to the Brock Ranch. Annie found a vertebra, and Jones and

133   THE NORTHERN PROVINCES

MISS ANNIE MONTAGUE ALEXANDER (1867–1950)

Annie Alexander's grandparents went to Hawaii as missionaries. Her mother, father, and another gentleman started sugar and pineapple plantations and became quite wealthy as owners of the California-Hawaii Sugar Company and Matson Shipping Lines. When Annie was fifteen her family moved to California.

In 1900 Annie attended John C. Merriam's lectures on paleontology at the University of California at Berkeley and became fascinated with his paleontological discoveries. On the condition that she would go along, she offered to financially support his summer expeditions. She loved camping and accompanied Merriam that summer to Fossil Lake, Oregon. This was the beginning of her close association with and support for paleontology at UC Berkeley. She financed much of Merriam's fieldwork from 1901 to 1907, including the early trips to Shasta County in 1902–1903 (which were the most successful in the history of California Mesozoic reptilian paleontology).

University of California Museum of Paleontology records show that Annie found twenty-seven Triassic reptile remains, as well as forty-three others with Eustace Furlong, for a total of seventy individual reptilian fossils.

Two species of Triassic reptiles were named for Alexander by Merriam, *Shastasaurus alexandrae* in 1902 and *Thalattosaurus alexandrae*

FIGURE 5.7

Miss Annie Alexander in the field. Courtesy of the University of California Museum of Paleontology.

Schaller each found two. On Thursday the twenty-sixth Osmont and Furlong rode horses to Brock's while Schaller and the ladies again looked for fossils. Annie found another rock full of bones, and they spent much of the day looking for more of that animal's remains. Jones found a vertebra.

From the twenty-sixth through the thirtieth they moved camp. On the thirtieth Miss Jones spotted a big rattlesnake on the trail, and Merriam, who had recently arrived in the field, took four shots with his pistol and killed it. It had eleven rattles, and Merriam commented that it was the largest he had ever seen. After putting the finishing touches on their new camp they immediately went to look for fossils. Annie found a small jaw with teeth, while Merriam found a rib, three vertebrae, and a limb bone. Around the campfire that night, after their long day of moving the camp plus lots of fossil hunting, Merriam entertained the group by telling "brilliant" stories.

in 1904. Annie had found the type of *T. alexandrae* in 1903. Merriam not only credits Alexander for her monetary support and discoveries but also mentions that she helped in the difficult preparation of some of the specimens. In 1906 she began making monthly financial contributions in support of paleontology at Berkeley.

In 1920, without informing Miss Alexander, Merriam left UC Berkeley for the Carnegie Institute in Washington. The remaining faculty in paleontology were merged, unhappily, with the geology department, prompting Professors Chester Stock and John P. Buwalda and Curator Eustace L. Furlong to leave for positions at the California Institute of Technology.

In 1921 Annie arranged to have her financial support used for the formation of the Museum of Paleontology at Berkeley, independent from the geology department. She established an endowment for the museum in 1934, and Merriam's successors continued to cultivate a close relationship with Alexander. In 1943 Samuel P. Welles named the plesiosaur *Hydrotherosaurus alexandrae* after Annie in honor of her continued support for the museum. In 1948 Annie Alexander gave more than sixteen thousand dollars of endowed scholarships to the museum. Annie was also very interested in mammals, birds, and botany and through her generous donations she was responsible for the establishment of the Museum of Vertebrate Zoology as well. Annie Alexander continued to show interest in mammalogy, ornithology, botany, and paleontology throughout her life, going to the field to collect whenever she could. She and her lifelong companion, Miss Louise Kellogg, spent her eighty-first birthday camped in the field.

In 2000 Garniss H. Curtis, professor emeritus of geology and geophysics at UC Berkeley, commented (pers. comm.): "She was a small woman, about five foot one, soft spoken, and the angel of paleontology at Berkeley." It seems that neither her sex, nor her height, nor her quiet voice deterred this giant in California paleontology.

The next day, July 1, they all went to where the previously discovered shastasaur lay in the rock. Osmont and Furlong got to work drilling holes in which to place the "giant" (blasting) powder. (This is the first written mention of black powder being used to extract reptilian fossils in California, a technique that continued into the 1940s in California and even later in Baja.)

Drilling holes for the powder using a hammer and bar is slow, hard work, but by noon they had "fired off two fuses." Afterward Merriam, Furlong, and even Alexander each drilled a hole. After setting too strong a charge, Merriam comments: "We blasted the holes and found nothing left of the saurian." So it goes for one California reptile that first lost its life and then—200 million years later—its legacy. That day, Jones discovered four "pavement teeth" and a vertebra, Alexander a tooth and a jaw, and Osmont and Merriam one vertebra each. In the evening they celebrated Jones's birthday with a venison dinner and stories around the fire.

On Wednesday, July 2, Merriam, Osmont, and Furlong went looking for a trail to another fossil site. On one steep section of the trail Merriam encountered bear hair, a subtle

FIGURE 5.8

Some bones prepared from the Hosselkus Limestone were distorted or cracked from the overlying weight of rock and the forces of plate tectonics. Here a shoulder bone from *Shastasaurus alexandrae* shows a crack completely filled with calcite. Courtesy of the University of California Museum of Paleontology; photo by the author.

signal of future events. That day Furlong managed to locate several fossil vertebrae, some ribs, and a saurian tooth, while Schaller found vertebrae and ribs from two individuals, plus several teeth. Alexander discovered numerous vertebrae, ribs, and jaws as well as three partial *Shastasaurus* remains and lots of individual bones. Jones found a tooth. With the large number of finds, this was a banner day in Triassic reptilian fossil hunting for California.

On Thursday morning it rained and all stayed in camp except Osmont and Furlong, who went to look for Pleistocene mammals in the limestone caves of the area. In the afternoon they all went to a new locality, where Merriam located a nice saurian embedded in the rock. Furlong found a saurian tooth, Alexander and Schaller a few bones, and Jones some vertebrae and other bones of an animal new to Merriam.

Friday was the Fourth of July, but not a work-free holiday here. Merriam and Furlong spent the day excavating Merriam's saurian remains while the others made additional discoveries. Miss Jones found a large vertebra and some other bones, while Osmont discovered a saurian jaw complete with teeth, plus some vertebrae and a small limb bone. Miss Alexander located a fine shastasaurian as well as three additional vertebrae. This was a happy Fourth for the field crew and another marvelous day for California reptilian paleontology.

On Saturday Osmont went for supplies, while Schaller had to depart the expedition. It was noon before the rest of the group went out to look at Alexander's saurian. As they were searching the outcrops for more specimens Alexander made a big discovery—this one with fur. Their first bear encounter prompted Jones to stay with Merriam and Furlong (who were working at removing one of the specimens found earlier) rather than venture out on her own. The bear did not deter Alexander, however. She continued to scour the slopes and was rewarded by finding another bone and, better yet, a shastasaurian complete with three of its four paddles! After finishing the job with Furlong, Merriam also managed to find more saurian remains.

All were abruptly awakened Sunday morning by Merriam firing his gun at an unsuspecting deer walking through camp. He simply rolled over where he lay and shot from his blankets. A crack shot he was not, for the deer ran off unhurt. At ten o'clock Furlong left for more fossil hunting. Busy breaking rocks, Furlong ignored a rustling in the bushes that he assumed was a deer. When he finally climbed a rock to see the cause of the ruckus, he found himself staring straight into the eyes of a bear cub. He hollered at the cub and it ran up the hill. As Furlong looked down from the rock, the huge cinnamon-colored mother bear, the largest he had ever seen, rose on her haunches above the brush and looked directly at him. The bear didn't threaten him, but Furlong avoided further shouting at the cub for fear its giant mother might attack. Instead he quietly collected his tools and slowly made his way back to camp. Once in camp he tried to convince Merriam that they should go back and kill the bear with his .38 caliber pistol. Merriam was smart enough not to oblige, saying that he was sure his wife and children wanted to see him again.

The morning of the seventh all left to collect the specimens. When they broke for lunch Merriam noticed smoke coming from the direction of camp. Osmont had failed to extinguish the breakfast fire completely, and the camp was on fire! Furlong immediately ran for camp, where he found the ladies' tent and all its contents burned to the ground and the fire still burning. With the foliage as dry as it was, the fire threatened to spread and consume the entire area. Jones reported that as Merriam took command he exclaimed "he would not let that fire get away from us for a thousand dollars," which was a lot of money in those days. He assigned everyone an area to control, and using sacks and tree limbs they fought the fire as best they could—though just as they seemed to gain the upper hand, the pine needles would flare up again. Nevertheless, by seven that evening they had finally extinguished the fire. Totally exhausted, they staggered back to camp and assessed the damage. The ladies had lost everything except their traveling dresses and coats that had been hung in a tree. Alexander's camera and film were lost, having been in the tent that burned. Among the destroyed photos were a picture of her saurian in the rock and several of the camp, a tragic loss of photographs from the expedition.

The provision tent survived, but with the ladies' tent gone, Osmont and Merriam bunked together and Furlong gave his bed to the women. He spent a restless night worrying that the fire might flare up again.

JOHN C. MERRIAM (1869–1945)

John C. Merriam stands out as the single most important vertebrate paleontologist in the history of Mesozoic reptilian discovery in California. Although he died at about the time I was born, I have grown to respect the man immensely for his work, expertise, and influence, which are woven throughout this history. In my fieldwork in paleontology in the west, he seemed to have beaten me to many sites, and because his list of publications is enormous, I am always bumping into his work.

Merriam first became interested in paleontology in his home state of Iowa, where he collected Paleozoic invertebrate fossils as a boy. He graduated from Lenox College and went on to the University of California, Berkeley, to study geology under Joseph Le Conte. At this time Merriam was an assistant in mineralogy. He then went to Munich to study Cretaceous mosasaurs under Karl von Zittel, where he received his doctorate in vertebrate paleontology in 1893. He returned to UC Berkeley the next year as an instructor in paleontology. In 1899 he made his first field expedition to Oregon's fossil-rich John Day country, and in that same year he was appointed assistant professor. He continued to build the vertebrate collections at Berkeley with material from northern California, the Mojave Desert, and Nevada.

In 1900 philanthropist Annie Alexander attended Merriam's lectures on paleontology and became fascinated with his discoveries. On the condition that she be invited to participate, she offered to provide financial support for his summer expeditions. The next year he took Alexander to outcrops in Shasta County where, in 1893, James Perrin Smith had stumbled on Triassic reptilian remains. This was the first time sci-

FIGURE 5.9

John C. Merriam in 1932. Charcoal on paper by artist Peter Van Valkenburgh. Courtesy of the University of California Museum of Paleontology.

The next morning, while the ladies searched for fossils in the burned area, Merriam and Furlong went to excavate the newly discovered saurians. The two men then took the afternoon off while Osmont and Alexander searched for more fossils.

The morning of the ninth they moved camp again. They started up the trail together on horseback, but when Annie's horse sat down and refused to move, Annie decided to walk. After a wonderful jaunt through the forest, complete with late-season wildflowers, they camped by a cool, clear mountain stream. They cut soft branches for their beds and told stories of their personal travels on into the night. The next day, while Osmont looked for fossils, the others stayed in camp. In the late afternoon Merriam and Alexander also went fossil hunting. Alexander came home about dinnertime, cold and wet to the waist: she had underestimated the depth of the creek.

Friday was a day for fishing and some fossil hunting. That afternoon Alexander en-

entists set out specifically to look for fossil Mesozoic reptiles in California. Over the course of the next decade Merriam's crews found a total of two hundred separate remains of Triassic reptiles in the Hosselkus Limestone of Shasta County (sixty-one of which are credited to Merriam personally). From these investigations Merriam was able to recognize a new order of Triassic reptiles, the Thalattosauria.

In 1905 Merriam was appointed associate professor at UC Berkeley. Three years later he published *Triassic Ichthyosauria, with Special Reference to the American Forms.* According to Chester Stock, "Probably no other single paleontological contribution by Merriam is as formidable and as valuable in its documented record of an extinct group of organisms as this work." Merriamosauria, a family of ichthyosaurs, is named for him.

From 1906 to 1913 Merriam collected a wealth of fossil vertebrates from Rancho La Brea in Los Angeles. In 1910 he became president of the Paleontological Society, and he was appointed full professor in 1912. In 1919 he became president of the Geological Society of America.

Merriam abruptly left Berkeley in 1920 to become the president of the Carnegie Institute in Washington. When Chester Stock and Eustace L. Furlong took positions at the California Institute of Technology, Merriam supported their work with funds from the Institute. Chester Stock said of Merriam, "He was by nature not genial, but rather grave and distant . . . [but he was] a polished speaker capable of holding and enthralling the audience with his subject matter."

Merriam's accomplishments are many. In reviewing his four volumes of publications and letters at UC Berkeley's Bancroft Library, I was overwhelmed at the number and variety of his interests and at his dedication to them. His interest in state and national parks, in the planting of trees in cities, and the preservation of redwood forests were particularly impressive. Merriam is well known for his work on fossil mammals in the western United States, but his work on California Mesozoic reptiles alone is enough to earn him a distinguished place in the history of American paleontology.

countered U.S. Geological Survey geologist J. S. Diller and his party while she was washing her hair in the stream. Jones related that Diller asked Alexander "all sorts of leading questions as to the plans of our party and in fact knew our movements as well as we did. She gave as evasive answers as possible." It seems Merriam's crew was leery of competition; indeed, many paleontologists (including, I have to admit, myself) are rather possessive of their favorite fossil hunting areas. Merriam returned that evening with fish for dinner, his pants wet and rolled up to the knees. The ladies found this a funny picture; as Jones commented, "[We thought] we would split our sides with laughter."

On Saturday they arrived at the next camp about noon. In spite of the heat, most of the group spent the afternoon looking for fossils. They discovered several bones of what Merriam thought was a *Nothosaurus,* and on the way back to camp they found even more. Osmont, who had gone fishing, returned with twenty-one fish, which, combined with hot cakes and soup, made for an eclectic evening meal. Merriam entertained the group with more of his stories around the evening campfire.

An interesting note comes from the paleontology collection at the National Museum in Washington, D.C. It is reported that U.S. Geological Survey paleontologist Timothy W. Stanton found the skull of a *Thalattosaurus shastensis* on this same day, July 11, 1902, in the Hosselkus Limestone of Shasta County: apparently Merriam's crew was not alone in the search that day. Merriam (1905b) mentions that Stanton loaned him this skull for study. According to Stanton's biography in *The National Cyclopaedia of American Biography,* vol. 40 (1955), he worked in the western United States and was interested in Mesozoic strata, his main paleontological interest being invertebrate fossils. During this same time he was also an instructor in paleontology and stratigraphic geology at George Washington University.

Sunday the twelfth, after nearly a month in the field, it was time to go home. Annie wanted to stay longer, but it just wasn't practical. On the way back, parts of the trail were steep, and Jones's horse was far from cooperative. Alexander would whip the horse from behind, and often Merriam had to tell Jones to dismount. As she attempted to remount, the horse would start before she was totally on. Jones commented, "Even the sedate Dr. Merriam had to laugh at my awkward attempt to get fairly on."

They returned to Oakland and Berkeley as they had come, by train, but after living in the wilderness for so long they found the change to civilization disquieting. That night, as they tried to sleep in their chairs on the rumbling train, Jones commented that they "missed the clear air and quietness of camp life." And so ended a wonderful and highly successful expedition, which we can better understand thanks to Miss Jones's delightful glimpse through a keyhole of paleontological history.

The next summer, 1903, Alexander financed and led another expedition to Shasta County. This time she was accompanied by Miss Edna Wemple, student of paleontology and one of Merriam's office staff. Wemple would become, in 1921, the first woman to earn a Master of Science degree in vertebrate paleontology from UC Berkeley (Stirton 1955). Eustace Furlong accompanied the two women to the field site, and the trio was later met by Merriam, Mr. F. S. Ray, and W. B. Esterly. This expedition proved its worth with the discovery of a specimen thought at the time to be a *Shastasaurus,* though after careful study by Merriam it became the type specimen of *Thalattosaurus alexandrae* (Zullo 1969). It was named for its discoverer, Annie Alexander.

I find no mention of a trip to Shasta County in 1904, and Merriam sent no expeditions to Shasta County in 1905 (letter from Merriam to Dr. A. Woodward Smith, May 25, 1906). In 1905 Alexander, accompanied by Wemple and Eustace Furlong, was off on a "saurian expedition" to look for Mesozoic reptiles in the Humboldt Range of Nevada (Zullo 1969). Although their work in Nevada distracted them from their efforts in the Klamath Mountains of California, it was quite successful.

FIGURE 5.10
This small jaw-like structure, originally from the Stanford collection, is called the pterygoid and is from the back upper pallet area of the skull of *Thalattosaurus perrini*. Courtesy of the California Academy of Sciences; photo by the author.

In a letter dated April 15, 1906, three days before the terrible earthquake and fire in San Francisco, Alexander sent Merriam a check for $800 to support that summer's fieldwork in Shasta County. On this expedition, Merriam hoped to find more thalattosaurian remains as well as to seek out Pleistocene fossils in the local caves. In the Hosselkus Limestone they were successful in finding bones from *Nectosaurus, Shastasaurus,* and *Thalattosaurus.*

James Perrin Smith returned to Shasta County in 1907, and on October 2 he wrote Merriam that in addition to his successful ammonite discoveries "I also found a jaw with teeth attached, that resembles *Thalattosaurus,* which I shall send or bring up to you."

In 1908 Merriam finished his paper entitled "Triassic Ichthyosauria, with Special Reference to the American Forms," in which he credits Furlong with "the largest share of the direction of the field work. . . . [He] prepared and mounted practically all of the specimens represented in the illustrations of the Californian materials represented in this paper."

At some point, J. S. Diller brought outcrops of the Hosselkus Limestone on Cow Creek (probably a branch now called Little Cow Creek) to Merriam's attention which had not been examined for reptilian remains. On May 23, 1910, therefore, Merriam, accompanied by paleontological assistant Bruce L. Clark (who had recently been appointed instructor in paleontology at Berkeley), R. W. Pack, E. L. Ickes, and Smith, was off to Shasta County once more (letter from Merriam to A. F. Lange, July 16, 1910). In addition, Clark brought three of his best students along. As usual they went by train to Bella Vista (letter from Merriam to Smith, May 21, 1910), where they hoped to get a team (horses with wagon) to go on to the limestones that same day.

In a letter dated June 6, 1910, to Alexander, Merriam suggests that their explorations had proved fruitful with the discovery of "scattered bones of *Thalattosaurus* and *Ichthyosaurus.* . . . We found a good many limb bones and vertebrae with a very few teeth. . . . Mr. Clark

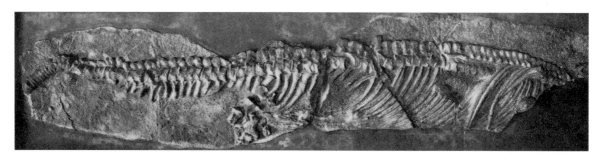

FIGURES 5.11 (TOP) AND 5.12 (BOTTOM)
Two rather complete skeletons found in Shasta County and pictured in
Merriam's 1908 monograph *Triassic Ichthyosauria, with Special Reference
to the American Forms.* Both photos depict *Californosaurus perrini,*
named for James Perrin Smith.

discovered . . . a very good specimen of *Shastasaurus,* which shows the finger portion of the
paddle. . . . Mr. Clark remained in the field to take the specimen out." Later, after excavation, Clark reported to Merriam that he had found the skull of *Shastasaurus* plus many other
"parts of the skeleton about which we know the least."

Although this expedition was very successful, it was not without its dangers. Merriam
was especially concerned about the abundance of mountain lions in the area. The "limestone region on Cow Creek in Shasta County seems to be a rendezvous for large cats. The
limestone caves are inhabited with a good many wild-cats, one of which came out and snarled
at us while we were collecting. While there I saw an old man who had just killed five mountain lions and was bringing the scalps in for the bounty" (letter to A. Alexander, June 6,
1910). (A personal note: I mapped geology in this same drainage for two years in the early
1970s [Hilton 1975]. I saw no sign of mountain lions and only one warm pile of bear ma-
nure. Sadly, it seems that hunting has severely reduced these animals' numbers.)

Nearly all the fossils brought back to Berkeley from these early expeditions were metic-
ulously prepared from the dense limestone by Eustace Furlong. He resigned before the summer expedition of 1910, however, and was temporarily replaced for that summer by Bruce

Clark (letter from Merriam to A. F. Lange, August 3, 1910). Clark was in turn replaced toward the end of summer by Pack and then that fall by B. T. Guintyllo, who, coming from the University of Kiev, did "not speak English very well" (letters from Merriam to A. Alexander, July 16 and October 3, 1910). Final preparation of the 1910 specimens was undertaken by a Mr. Rope and Loye H. Miller, who worked on the skulls (letter from Merriam to Bruce Clark, June 6, 1910). Miller, a 1900 graduate of the University of California, had been recommended by Merriam to be hired as instructor of paleontology (letter from Merriam to A. F. Lange, July 16, 1910).

Artists involved in illustrations of Merriam's original works include Raymond Carter, Mrs. Grace Ballantine, and A. J. Heindl. Photographic credit for some of Merriam's early work goes to B. F. White.

Welles: 1949

The next time UC Berkeley fossil hunters went to the Shasta County area was nearly forty years later, when in 1949 Samuel Welles took a group to some of Merriam's old haunts. From Welles's field notes of Tuesday, August 9, 1949, we have this account: "Left Lair of the Golden Bear [the Berkeley campus] at 6 A.M. with 3 cars, incl. Mr. Sibley, Dr. Blair & son, Mr. Bach, Miss Bolte, Miss Flynn, Mr. & Mrs. Brown, Mr. Palmer & sons of Redding & 3 Welles & Andy." When they reached the site in Shasta County they

> parked cars and hiked up road evidently built by Merriam's parties. . . . About 100 feet from car park I found a ls [limestone] block with impression of skull side with blunt teeth about 1/2 inch long. Little left of skull but it shows lateral portions of 4 posterior teeth and possibly more anterior. . . . After lunch took left side of ls. cliff and about 100 yds along lower edge I found a mass of ribs probably representing belly ribs of a small ichthyosaur. Made no attempt to excavate as Palmer will return and have more time to do a careful job. I was afraid of ruining what might be a fairly complete skel. [skeleton]. Don't know whether head is in the bank or out!

No definite reference to this find is made in the UCMP collection, although there are several undated finds simply stating that they were made by a "U.C. Party."

Recent Expeditions: 1972 to the Present

In the summers and weekends of 1972 through early 1974 I mapped the Hosselkus Limestone of the Little Cow Creek drainage for my master's degree project. I found several fragmentary reptilian remains, with one site having many bones still embedded in the limestone,

FIGURE 5.13

A Late Cretaceous Sierran river scene, with the hadrosaur *Saurolophus* accompanied by the toothed bird *Ichthyornis,* a modern shore bird, and the pterosaur *Pteranodon.*

where they remain still. Years later as a college instructor I took some students back into the area and managed to retrieve a couple of nice vertebrae.

New people at Berkeley have continued the search in the Hosselkus. In 1997 ichthyosaur specialist Ryosuke Motani and turtle specialist James Parham, along with geologist Pat Embree and paleontologist Eric Göhre, visited the Hosselkus Limestone area where I had found the bones in the early 1970s. They brought back a few fragmentary specimens. In 1998 they were off again, looking for some of Merriam's areas, but so far they haven't had luck finding significant new specimens. Pat Embree did work out a couple of bones at the Little Cow Creek site, which he graciously donated to the Sierra College Natural History Museum collection.

With more hard work I am sure there are many additional discoveries to be made in the Hosselkus Limestone. Unfortunately, because limestone weathers so slowly, most new finds will probably have to be made in unsearched or poorly searched areas.

SIERRA NEVADA PROVINCE

In the Sierra foothills along I-80 there is a legend of a creature having been found that locals call the "Clipper Gap Monster." A newspaper first reported the find in 1902, and the myth persists to this day. It seems the monster had lain (unnoticed?) for decades in a cut bank of the Central Pacific Railroad. I have been to the site, and the rocks here are metamorphic and a weathered mess—not rocks likely to have preserved a fossil of any kind. The article reproduced here (see fig. 5.14) mentions "scientific men," but from what institution? What type of petrified skeleton? And where did it go? Although many in the central Sierra Nevada foothill country still believe in this creature, and some think they can still see it in the cut, this first report of a possible reptilian creature in the Sierra Nevada is pure folklore.

The Sierra Nevada is in fact one of the most unlikely places to find Mesozoic reptilian remains in California. These rocks have been squeezed, deformed, and heated to such an extent that most fossils simply could not survive. Also, many of the rocks in the Sierra originated in deep offshore areas where reptilian life would not have been abundant. Although the Hosselkus Limestone that is so rich in reptilian fossils in the Klamath is also found in the northern Sierra Nevada, thus far no Triassic marine vertebrates have been found in it. Nevertheless, two major exceptions to this paucity of fossil remains do exist, one from the middle of the twentieth century, the second from just a few years ago.

Another Petrified Skeleton.

Under the heading of "A Monster in the Rocks" the last issue of the Sacramento Sunday News contained the following story: "A party of scientific men have been engaged for a week or more in cutting out of the solid rocks at Clipper Gap, on the line of the Central Pacific Railroad, the remains of a monster belonging to some extinct species. The skeleton is petrified and is about twenty-five feet long. The exposure was first indicated when the railroad builders cut into the rock to make a roadbed for the tracks. The rock is of a yellow color but the petrified animal is nearly black. The outlines of the strange creature are quite distinct, although it must have been confined in its prison house in the rocks for thousands of years. The scientific party attach great importance to the discovery. Prior to its removal the skeleton was plainly discernible from the passing trains, but nobody cared to inquire into the mystery. The creature was evidently of amphibious habits and its presence in the mountains confirms the theory that a sea once rolled over the greater part of California."

An Auburnite makes the suggestion that the Grass Valley owner of the "petrified man" unearthed in this locality some time ago, purchase the petrified monster at Clipper Gap, and a remunerative dime-drawing museum will be the result. Probably the Cleopatra's Needle on the Thames embankment can be secured to make the show all the more captivating.

FIGURE 5.14

Reprint from the January 22, 1902, *Republican Argus* of Auburn, California: "A Monster in the Rocks, Strange Discovery by Scientists Up at Clipper Gap" (originally published in the *Sacramento Sunday News* three days earlier).

Clark and Welles: 1954

The first discovery of reptilian fossils in the Sierra was made in 1954 by Lorin D. Clark of the U.S. Geological Survey. These remains, found in a large limestone concretion on the west bank of McClure Reservoir in Mariposa County, were those of a plesiosaur, and they were lodged in the Late Jurassic Salt Spring Slate, the probable equivalent of the Mariposa Formation. Sam Welles, an expert on plesiosaurs, was taken to the site on December 11, 1954, together with his paleontology class. The following are excerpts from his field notes:

Sat 12/11/54 . . . out to McClure reservoir . . . [with] Party of Paleo III [students], Joanne Klein, Tom Rogers, Bill Haney, Marilyn & Allan Seagrave. Met Loren Clark, U.S.G.S. . . . hiked up over hill to tiny ls [limestone] lens at high water level in slates. . . . Three small ls.

FIGURE 5.15
Plesiosaur bones found by
Lorin Clark and prepared
for publication. Courtesy of
the University of California
Museum of Paleontology;
photo by the author.

lenses are the only ones in the whole region and the reptile is in center lens. Collected a number of small bone-bearing fragments. . . .

Sun 12/12/54 Nearly froze last night. . . . collected slab from top of ls. lens. . . . located on map, took photos & back to Berkeley.

He went back to the site in January accompanied by Professor Taliaferro and Pete Norton to try to find more pieces. Here he writes:

1/25/55 . . . Cold and foggy 'till 1:00 & then hot. Found other pieces to contact skel. [go with skeleton]. . . . Still need one piece in center of block and the distal piece of a propodial . . . so some is still missing and I presume that a thorough, 2-day search would turn up the parts.

Welles was working on a publication on this find at the time of his death, August 6, 1997. He concluded that the creature was one of the short-necked plesiosaurs from the family resembling the Cryptoclididae.

Christe: 1999

The second Sierra find came in 1999, when paleontologist Kevin Padian of the University of California Museum of Paleontology forwarded a letter he had received from geologist Geoff Christe, a specialist in the Late Jurassic rocks of the northern Sierra Nevada. Christe enclosed

FIGURE 5.16
A hypsilophodontid, a deer-sized herbivorous dinosaur, takes a drink
from a stream in an Early Cretaceous northern California forest.

pictures of Late Jurassic fossil bones that he had found in the upper Feather River country of the Sierra, in rocks he interprets to be "fossil" river channel deposits. Finally, we have terrestrial deposits in California that are yielding dinosaur bones! Padian was kind enough to suggest that I work with Christe in the study of the bones and to help search for others in the area. The bones comprise a few fragments, a possible but as yet unidentifiable tooth, the unmistakable proximal end of a rib, and a possible neural spine from a vertebra. The rib appears to be from a dinosaur with a rib cage about the size of that of a modern bison (Christe and Hilton 2001).

Christe made the discoveries in steep terrain in October of 1995 and 1997 while doing geologic fieldwork. They were found in the highly inclined, hard, but lightly metamorphosed Late Jurassic rocks of the Trail Formation. He recognized their importance as possible Jurassic dinosaur fossils because he had previously discovered dinosaur footprints in Manassas, Virginia (Weishampel and Young 1996).

GREAT VALLEY PROVINCE: SACRAMENTO VALLEY

The Sacramento Valley is the northern half of the four-hundred-mile-long Great Valley of California. Around its edges outcrops of rocks from the Great Valley Group have yielded several fossil Mesozoic reptile remains. These rocks range in age from very latest Jurassic (ca. 150 MYBP) to very latest Cretaceous (ca. 65 MYBP). Most of the outcrops are exposed in linear hills where uptilted edges of marine sedimentary rocks form hogbacks and strike valleys up against the northern Coast Ranges on the west side of the valley.

Other outcrops occur along the east side of the Valley, where stream erosion has cut through the younger volcanic rocks of the Cascade Province or where the Sierra Nevada Province dips beneath the Valley sediments. The Sacramento Valley receives more rainfall than the San Joaquin Valley to the south, so there is more erosion but also more plant growth. Good exposures are therefore harder to find than in drier areas such as the Panoche Hills of the San Joaquin.

Early Discoveries: 1908 to 1914

The first mention of Mesozoic reptile remains discovered in the Sacramento Valley area is in a letter by John C. Merriam dated March 17, 1908. In a 1910 letter to C. K. Studley of the California State Normal School at Chico (now CSU Chico), Merriam refers to this specimen as "a single reptile tooth with two fragments of bones from the Chico Formation" and as the only reptilian remains found thus far in the region. At the time, *Chico Formation* was a term used for most of the rocks that are today considered Great Valley Group. He mentions

FIGURE 5.17
Hogback ridges of the Great
Valley Group on the west
side of the Sacramento
Valley; view looking east
with East Park Reservoir in
center. Photo by the author.

nothing else about this find, and the specimen does not appear in any of the collections. (Studley had come to Chico Normal School in 1907. He later became the head of the geology department and in 1931 became dean of the college and president pro tem [*Record* (Associated Students of Chico State College) 34 (1932)].)

The first Jurassic reptilian evidence for this part of California comes in three letters exchanged in 1913–1914 between Merriam and Thomas L. Knock, a civil engineer in Willows, California. Knock, who had some training in geology, describes a fossil vertebrate that he thought was an ichthyosaur (but was most likely a plesiosaur). He recounts, "The vertebra[e] are about 5 inches in diameter, then there are a lot of small bone[s], rib bone[s] apparently. . . . The whole thing [specimen] was some thirty or forty feet long." He also enclosed a map of its location near Stony Creek between the towns of Stonyford and Elk Creek. He accurately described the site in township and range parameters and included pictures of some of the bones. According to the map, the fossil would be from the marine Great Valley Group and of latest Jurassic age. The pictures he provided to Merriam clearly show large vertebrae that are concave in their centrum as in some Late Jurassic plesiosaurs. Merriam's last letter to Knock asks if the university could keep a single vertebra that Knock had sent to him. What happened to that vertebra and the rest of the remains is not known, but it seems that Merriam lost an important opportunity to collect the most complete Jurassic reptile specimen ever found in California.

FIGURE 5.18

Patty and Clifford, the
daughter and grandson of
Inez Kelly (who found the
specimen), proudly display
the "fossil horse hoof"
(actually a plesiosaur vertebra) they were using as a
doorstop. Photo by
the author.

Tehama County Discoveries: 1950 to 1960

Several finds were made on the west side of the Sacramento Valley in Tehama County between 1950 and 1960. In about 1950, Mrs. Inez Kelly, a rancher in the northern Sacramento Valley, happened upon what she thought looked like a fossil horse hoof. When she recently showed it to me she was using it as a doorstop outside her front door. I informed her that it was actually a plesiosaur vertebra, and she generously donated it to the Sierra College Natural History Museum collection (see fig. 5.18). In 1954 Bates McKee, assistant professor at the University of Washington, found a vertebra and a flipper bone of what is identified in the UCMP collection as a pliosaur, one of the short-necked plesiosaurs. There is a report of a G. Young finding a small, unidentifiable fragment of bone in 1957, which is still in the collection of the UCMP. Rancher Donna Rae Wahl-Hall found a possible plesiosaur vertebra in "about" 1959 and "loaned it to a professor"—possibly McKee?—who she said gave it to the UCMP collection. A news article in the *Sacramento Bee* on April 30, 1960, names McKee as having found a plesiosaur bone "20 miles west of Corning [Tehama County]." This discovery is further problematized by the fact that Welles (n.d.) mentions McKee finding an isolated posterior cervical vertebra of a plesiosaur from the genus *Trinacromerum* on the property of Mr. Karl Wahl in the "summer" of 1960. The coordinates for this specimen in the Berkeley collection do not match the spot where Mrs. Wahl-Hall said she found hers. Whether one or two or perhaps even three vertebra were found here is unclear; in any event, only one is inventoried.

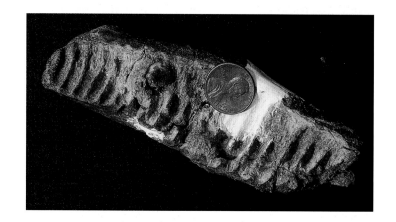

FIGURE 5.19
Hadrosaur maxilla discovered by Gerard Case in the Late? Cretaceous Great Valley Group on the west side of the Sacramento Valley. Courtesy of the Yale Peabody Museum; photo by the author.

Case: Early 1960s

While I was inquiring about possible Triassic reptile fossils in the Yale Peabody Museum collection, I stumbled upon an exciting rediscovery important to northern California paleontology. Briefly mentioned by Horner (1979), it seems that Gerard R. Case, a prominent paleoichthyologist (one who studies fossil fish), discovered most of the left maxilla (upper jaw) of a hadrosaur in Tehama County in the western portion of the Sacramento Valley. The fossil is listed as Late Cretaceous, as are all previous hadrosaur remains from the western slope of North America; however, the geologic map of the area shows this find to have come from Early (lower) Cretaceous rocks. This is the first and only record of hadrosaur remains found in California north of the San Joaquin Valley. According to Case (pers. comm. 1998), between 1962 and 1964 he was in Tehama County searching for Late Cretaceous shark teeth when he found the dinosaur remains in a road cut. He donated the specimen to Princeton when he moved to the East Coast. Princeton later transferred its collection to Yale, where the fossil is currently housed.

A Few Finds in the 1970s

In the mid-1970s Barbara Hail, who runs a small antique shop in the town of Elk Creek, found a fossil vertebra in Elk Creek (the creek) between Stonyford Dam and the town. From her description (pers. comm. 1999) it sounds like a plesiosaur vertebra. This is fairly likely because a plesiosaur vertebra was discovered just upstream in a gravelly boulder by Justine Smith in 1998, and Ken Kirkland found a portion of a plesiosaur humerus in the area in 1999. Mrs. Hail said she loaned her specimen to a friend who took it "back east" for identification and may have left it with a university in Massachusetts.

In 1976 Peter Rodda of the California Academy of Sciences and Michael Murphy of the

FIGURE 5.20
Ammonite with possible
reptile bite marks (on both
sides of the shell) found by
Peter Rodda and Michael
Murphy in Cretaceous rocks
of the Great Valley Group.
Courtesy of the California
Academy of Sciences; photo
by the author.

FIGURE 5.21
Ammonite found by Bill
Spangler (right) and the
author from the Great Valley
Group in the Sacramento
Valley. Photo by Judith
Peyret; from the collection
of the author.

University of California at Davis found what may be tooth marks of a reptile in the shell of an ammonite in Shasta County. Plesiosaurs and mosasaurs also lived during the Cretaceous and typically ate ammonites.

In 1995 Rodda found a second ammonite shell in Shasta County with apparent bite marks from a reptile. Recent work by T. Kase et al. (1998) suggests that similar marks found on other ammonites may instead be limpet (a mollusk) attachment marks, so whether the specimens mentioned here truly have bite marks cannot be said with certainty.

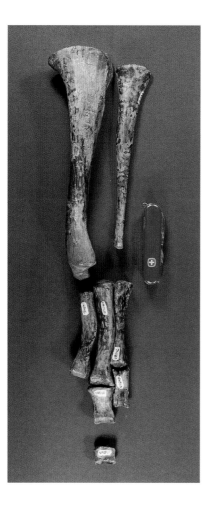

FIGURE 5.22 (LEFT)
Leg of hypsilophodontid
from Shasta County. Photo
by the author.

FIGURE 5.23 (RIGHT)
Pat Embree doing final
preparation on a hyp-
silophodontid bone found
in Shasta County. Photo
by the author.

Hilton: 1988 and 1991

In 1988 I found a small costal (expanded rib) of a fossil turtle (Chelonia) in the Late Cretaceous Chico Formation in Mill Creek Canyon on the east side of the Sacramento Valley.

In early December 1991 I was looking for ammonites in Shasta County accompanied by
my son Jakob and geologist Tom Peltier when I found two concretions containing bones in
Early Cretaceous siltstones. It wasn't until two years later, when fossil preparator Pat Embree
worked the bones from the concretion, that we realized their importance. There were eight
bones in all, representing most of the left hind limb (from the knee down) of a deer-sized,
fleet-footed, herbivorous hypsilophodontid dinosaur.

In 1998 a mounted cast of a skeleton of a hypsilophodontid (originally found in Canada)
was assembled by paleontologist Frank DeCourten of Sierra College. A cast of the Shasta
leg was also put on display for comparison. DeCourten and Sierra College students built a

FIGURE 5.24
Frank DeCourten displays the completed cast of the skeleton of a hyp-silophodontid assembled for the Sierra College Natural History Museum.

display to house the skeleton, where it can be seen in Sewell Hall at the Natural History Museum at Sierra College in Rocklin, California.

Antuzzi and Göhre: 1990s

The first theropod (meat-eating dinosaur) remains found in California were found in 1994 by fireman Patrick Antuzzi in Granite Bay near Sacramento. Pat found the midsection of a long bone while searching a ditch excavation in a subdivision. Microscope analysis of the bone by Greg Erickson (then at the University of California, Berkeley) indicated it was a theropod. Antuzzi also discovered skull fragments and a tibia of the first mosasaur to be found in the Chico Formation that same year (Hilton and Antuzzi 1997). According to Gorden Bell of the South Dakota School of Mines and Technology (pers. comm. 1999), this mosasaur may have affinities with forms found in New Zealand. From 1994 to 1996 Pat and I worked closely in the same area, finding several other reptile bone fragments as well as the fragmentary remains of two types of turtle from the families Cheloniidae and Dermochelyidae (Parham and Stidham 1999).

In 1997 I was introduced to Eric Göhre by Patrick Embree. Eric had been collecting in the Chico Formation on the east side of the Sacramento Valley for years. Among the beautiful fossil shells, shark teeth, and leaves that he had found, he also had some bones, two of which appeared to be bird bones. The bones were still obscured by the rock matrix. If they really were bird bones, these would be the first Mesozoic fossil bird remains from California. I borrowed

PATRICK J. ANTUZZI (1964–)

Patrick Antuzzi has been collecting fossils since the age of six, when he found his first one in the sea cliffs near Santa Cruz, California. He has since collected fossils in many areas of California as well as in Nevada, Arizona, Montana, Utah, Wyoming, and Missouri. He takes his hobbies very seriously. He is an expert fly fisherman. He also raises exotic birds, is an excellent chef, a serious gardener, and excels in photography. Today, Antuzzi makes his living as a fireman, and in doing so has won several awards, one of which was for integrity.

His love of fossil hunting paid off for him in 1995 when he

FIGURE 5.25
Pat Antuzzi assembles mosasaur skull fragments found in the Chico
Formation in Granite Bay. Photo by the author.

ERIC GÖHRE (1959–)

Eric Göhre is one of the most important collectors of Mesozoic reptilian fossils in California. He started fossil hunting while a senior in high school, when he decided to do a project on fossil mollusks from the Chico Formation. He has been collecting in this formation in the hills on the east side of the Sacramento Valley ever since. After high school he went on to California State University, Chico, earning a bachelor's degree in geology in 1984. Over the years Göhre has collected fossils in other parts of California as well as Nevada, Utah, Oregon, Idaho, Wyoming, and Colorado.

His personal catalogued collection of fossils, which numbers in the thousands, has been used in several papers published on the Chico Formation, one of which he co-authored. He hopes that one day his sub-

FIGURE 5.26
Eric Göhre inspects a fossil at one of his favorite collecting sites
in Butte County. Photo by the author.

those that needed preparation and took them to Sierra College, where fossil preparator David Maloney tediously worked them out of their enclosing matrix. Then Eric and I took the bones to the UCMP, where we talked to three key people who gave us very interesting answers.

Pterosaur expert Kevin Padian told us that one of the bird bones was actually a pterosaur wing bone. A previous account of pterosaur remains was made by Downs (1968) from the discovered the partial remains of the first theropod dinosaur to be found in California. His other discoveries include the fossil remains of an undescribed new species of dermochelyid turtle, and one of the oldest cheloniid turtles to be found on the west coast of North America. He also found a partial skull of a mosasaur from the genus *Clidastes,* not previously recognized in North America: the first known mosasaur from the Chico Formation.

These important discoveries led him to further his education. He has taken several geology courses and in the summer of 1999 he went on a Sierra College field course in search of dinosaurs in Montana. His uncanny knack for finding fossils came through there, too, when he found fossil turtle remains, the partial remains of a hadrosaur, and a *Triceratops* brow horn. Antuzzi most assuredly will continue to make important contributions to paleontology.

stantial collection and more extensive mapping of the Chico Formation will be the basis for a more thorough scientific study of the Late Cretaceous paleoenvironment in northern California.

Göhre's most important fossil finds, all from the Chico Formation, include California's only Mesozoic bird remains: an *Ichthyornis,* a *Hesperornis,* and a neognath. He has also found bones from two pterosaurs, among the first to be scientifically documented from California. He found the first mosasaur remains in California north of Sacramento as well as important Mesozoic turtle remains. He has graciously donated many of his reptile fossils to the University of California Museum of Paleontology and the Sierra College Natural History Museum, where they are available for scientific study.

Göhre works as a paleontologist for an environmental paleontological monitoring firm. A talented artist, he enjoys drawing his fossils and recreating what these animals may have looked like when they were alive. In years to come Göhre will undoubtedly make other important contributions to paleontology.

Late Cretaceous Moreno Formation in the Panoche Hills in the western San Joaquin Valley of California, but that specimen cannot be located. Thomas Stidham, a specialist in fossil birds, informed us that we also had a humerus from the toothed bird *Ichthyornis.* The other bones, identified by fossil turtle expert Jim Parham, turned out to be from Late Cretaceous turtles, and two vertebrae were from mosasaurs.

In 1998 Eric Göhre took Pat Embree, Tom Stidham, Jim Parham, Pat Antuzzi, and me to his fossil sites. Göhre shortly afterward found a huge wing bone (fourth metacarpal) of a

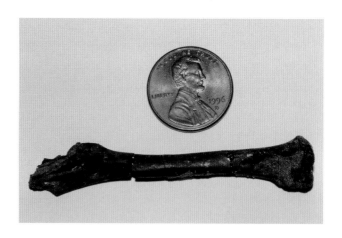

FIGURE 5.27
Ichthyornis wing bone (humerus) found by Eric Göhre in the Chico Formation: the first Mesozoic bird bone from the state. Courtesy of the University of California Museum of Paleontology; photo by the author.

FIGURE 5.28
Pterosaur expert Wann Langston holding Göhre's wing bone from a pterosaur from the Chico Formation. In life the pterosaur had a wingspan of about eighteen feet. Photo by the author.

pterosaur that, when alive, would have had about an eighteen-foot wingspan. He also later found an ulna from an early representative of living birds (a neognath). Up in Shasta County, Embree was looking for ammonites and found a fragment of bone from the oldest pterosaur found on the West Coast. It is Early Cretaceous and from the same area where the hypsilophodont dinosaur was discovered.

FIGURE 5.29
Rancher Jim Jensen Jr.
proudly displays a portion of
a plesiosaur humerus found
on his ranch. Photo by the
author.

Hilton, Maitia, and Others: 1998

While working with Göhre on the east side of the Sacramento Valley I enlisted the help of
logger Doug Maitia and others to help collect on the west side of the valley in the older Jurassic rocks of Tehama County. After interviewing Donna Rae Wahl-Hall and several other
ranchers in the area, I realized that this region had the potential to yield other rare Jurassic
reptilian remains. Jim Jensen Jr., another local rancher, had collected the large midsection of
a long bone from a plesiosaur in 1995; Dave Mattison, a geology instructor at Butte College,
had found a plesiosaur vertebra in the same area; and there were rumors of other finds.

The hilly area on the west side of the Sacramento Valley is still sparsely populated and
very wild. On our subsequent fossil collecting trips we encountered wildlife not often seen
in California today, including three species of owls, numerous hawks, falcons, osprey, and a
few golden and bald eagles. Deer were plentiful, rattlesnakes and other snakes fairly common, and coyotes (other than those hung by ranchers on fences) were often seen. We saw a
few feral pigs, a goat, and even a badger trying to kill a skunk.

FIGURE 5.30
Logger and fossil hunter
Doug Maitia. Photo by
the author.

On the first trip on June 13, 1998, we went to a local ranch where Maitia had heard a rumor of a "dinosaur skeleton" embedded in the rock along a local ridge. The party consisted of Maitia, preparator Dave Maloney, a young boy named Clifford May, and myself. The hike was grueling and steep, and although we didn't discover any dinosaur remains, we did find other fossil bones: a small tail vertebra from a plesiosaur still embedded in the rock, a small fragment of a distal limb bone that had already weathered out, and a third large bone still embedded high up on an overhanging cliff (and that we would have to wait for another day to collect). The last bone discovered that day was the proximal end of a scapula that Maitia found embedded in very hard rock. It took chisels to extricate this one. On that single day we doubled the collection of Jurassic reptile fossils found in California, and they were all from plesiosaurs.

On September 26 Maitia, rancher Jim Jensen III (Jim Jensen Jr.'s son), and I prospected on another ridge to the north. In a rather small area we found a rib fragment, a scapula, and two neural spines from vertebrae. The scapula and one of the neural spines proved to be from Jurassic plesiosaurs.

FIGURE 5.31
At the top of the ladder, and still the ischium (hip bone) is well above me. Photo by Vicki Van Why.

The next month, on October 25, I went back to the first site, accompanied by fossil hunters Eric Göhre and Vicki Van Why, to retrieve the large bone in the overhanging cliff. It had rained a bit, and the rancher didn't want us to drive on his dirt roads for fear we might damage them. The rancher was, however, willing to give us the key to a gated gravel road that would get us within (arduous) hiking distance of the site.

With a second permission to cross another property and a warning (perhaps to discourage us) of "rabid mountain lions," we were on our way. We had to carry climbing ropes, a harness, and a large ladder up steep brushy slopes for nearly two miles. At the overhanging cliff we tied the ladder in from above, while I got fastened into the climbing gear at the base of the cliff. Van Why and Göhre held the two safety ropes from above while I climbed the cliff and then made my way to the top of the suspended ladder, where I could just get at the fossil about twenty-five feet up. I impregnated the bone with Vinac (a liquid plastic) and gave it a few minutes to dry. I then covered the bone with modeling clay so that pieces wouldn't be lost if they broke off. It took about half an hour of chiseling, the rock chips raining in my face, but we finally got the specimen. On the way down the ladder I lost my footing and found myself dangling in midair, suspended from the ropes held firmly by my now dearer friends Vicki and Eric. The bone turned out to be the ischium of a Jurassic plesiosaur.

The following week Ryan Warren brought me a fossil vertebra that his friend Justine Smith had found in a gravelly boulder upstream from the town of Elk Creek, nearly thirty miles south of the area we had been searching. It was a Jurassic plesiosaur vertebra, and this find gave us hope that there were many more fossils to be found.

FIGURE 5.32
Strange branching gastralia
(belly rib) from a plesiosaur
found by Paul Hilton. Photo
by the author.

On November 11 I set out again for Tehama County and met Maitia and my brother Paul Hilton near the site. We found a neural spine fragment, and Paul noticed a strange bone that resembled a small branching deer antler. It turned out to be a gastralia (belly rib) of a plesiosaur.

On December 5 we were up that way again. After a day of fruitless hunting that ended in a snowstorm, rancher Jim Jensen III gave us some fragments from a small, dense bone he had found on the ranch. After assembling the bone I showed it to Mark Goodwin at UCMP, who concurred with my suspicion that it might be part of a dinosaur foot. I sent it to John Horner at the Museum of the Rockies in Bozeman, Montana. Horner thought it was a saurischian dinosaur metatarsal, but couldn't say for sure which kind. I then sent it to Philip Currie at the Royal Tyrrell Museum of Paleontology in Alberta, Canada, but although he agreed it was dinosaurian, he couldn't say what kind it was either. He showed it to Donald Brinkman and Elizabeth Nicholls, who ruled out plesiosaur and turtle, and Wu Xiao-Chun, who said he didn't think it was a crocodilian. I showed it to Ryosuke Motani at the University of California Museum of Paleontology, and he ruled out ichthyosaur as well.

The day after Christmas I had arranged to show veteran fossil hunter Chad Staebler (see San Joaquin section, p. 216) where we were getting the Jurassic plesiosaur remains in the northern Sacramento Valley. Shortly before this I got a call from my eighty-year-old friend Allan Bennison, who at seventeen had found the first dinosaur in California. He said he would be in the area for Christmas and asked me if I would like to go fossil hunting the day after. So Staebler, Bennison, and I went together, and we met up with Maitia as well. I had obtained a permit from the Bureau of Land Management to work a ridge where Butte College

FIGURE 5.33 (ABOVE)
Possible dinosaur
metatarsal (proximal end)
found by Jim Jensen III.
Photo by the author.

FIGURE 5.34 (LEFT)
Left to right: Allan
Bennison, Dick Hilton,
and Chad Staebler. Photo
by Doug Maitia.

geology instructor Dave Mattison had found a plesiosaur vertebra. It was a steep, eight-hundred-foot ridge, but I wasn't concerned about Bennison because I knew that even at his age he could out-hike any one of us.

We got lucky early in the day, when just below a cliff at the summit area Maitia found a fossil midflipper bone from a plesiosaur. Although we continued the search, the hills yielded no more fossils that day. Still, we had a good time, survived a run-in with a nasty feral billy goat, and we exchanged many stories of past success.

Sierra College Volunteers: 1999 and 2000

By 1999 we had a whole group of Sierra College Natural History Museum volunteers searching for Mesozoic reptiles in northern California. On January 23, 1999, geology student Sue

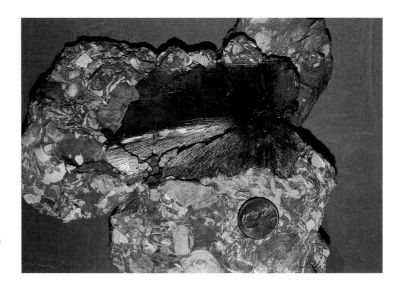

FIGURE 5.35
Part of the carapace of a
Late Cretaceous turtle
from the Chico Formation
discovered by Eric Göhre.
Photo by the author.

Gardner and I met with Eric Göhre in the area of Butte County where he had found his
fossil Cretaceous bird and pterosaur remains. Our purpose was to do some reconnaissance
toward getting a detailed stratigraphy of the area, which Gardner and fellow student Mandy
Lauenroth (now both geologists) would render in map form. We were successful in prelim-
inarily tying the three main outcrops in the area together stratigraphically. We also came across
part of a fossil turtle shell and a fragment of a pterosaur bone. Göhre found partial remains
of several turtles and the vertebra of a mosasaur.

On March 6 Maitia and I searched more of the Tehama County area where Maitia had
discovered the far end of a plesiosaur rib. On May 15 I went in search of reptilian remains
in Glenn County, accompanied by Sierra College biology professor Charles Dailey and fossil enthusiasts Vicki Van Why, Pat Antuzzi, and Ken Kirkland. Our goal this beautiful spring
day was to comb more of the Jurassic beds south of Elk Creek. En route we looked for the
site where Thomas Knock had found the thirty- to forty-foot remains of a plesiosaur(?) in
Jurassic beds between Stonyford and Elk Creek. Having no luck in finding any more bone
material, we decided to try again on our way home.

In Elk Creek after just half an hour of climbing Kirkland found part of a large bone sticking out of a huge boulder. It was the distal end of the left humerus of a plesiosaur. On the
return trip we stopped again at the Knock site, where Antuzzi found the possible mold of a
bone about the size and shape of a plesiosaur forearm.

On April 9, 2000, I met with Bennison to do fieldwork in the same Late Jurassic deposits we had been searching for a couple of years. Even at eighty-two years of age, he

FIGURE 5.36

A dead Late Cretaceous plesiosaur being scavenged by a mosasaur and
sharks while a large sea turtle escapes the scene.

could still hike the rugged gravelly hogbacks of Glenn County. The plesiosaur material we had previously found had just been returned to me by Glenn Storrs of the Cincinnati Museum Center, and although he knew we had both a long- and short-necked plesiosaur, he couldn't tell us what genus they were. Bennison and I were out looking for more diagnostic material.

We were able to get permission to explore about a quarter mile of ridge. This wasn't much territory, but between my hay fever and Allan's age it turned out to be just right. Finding nothing, we decided to drive south and check the road cuts. When we came to where Stony Creek cuts through the ridge we stopped to explore the gravel. Allan had more energy than I did and decided to go down to where the creek cut through the gravelly layers. Here Allan exclaimed that he thought he had spotted a bone. He pointed out a cream-colored circular object about the size of a pea, with a round hole in it. I was pretty sure it was a belemnite (the skeleton of a squidlike creature), though if he was really lucky it might be a tooth.

The next day I began to prepare the belemnite/tooth out of the pebbly conglomerate. To my amazement, it proved to be a long and beautiful sharp-ridged plesiosaur tooth. So Bennison has now found the first dinosaur, the best mosasaur skull, and the first Jurassic plesiosaur tooth in California!

There is much left to do in the search for reptilian remains around the hilly edges of the Sacramento Valley. With the help of local ranchers, we should be able to uncover many important new fossils.

GREAT VALLEY PROVINCE: SAN JOAQUIN VALLEY

The San Joaquin Valley occupies the southern half of the Great Central Valley of California. The hills on the west side of the San Joaquin Valley contain Late Cretaceous rocks of the Great Valley Group that have yielded a treasure trove of reptile fossils. These fossils were deposited offshore from the ancestral Sierra Nevada in a marine environment between about 80 and 65 million years ago. These rocks also yield the skeletons of fish, invertebrate marine creatures, and plant remains. The plant remains (wood, leaves, needles, and fern fronds) and even dinosaur carcasses washed down rivers into the sea. After becoming waterlogged they settled to the bottom and were sometimes preserved. Marine reptiles perished here too. Marine reptiles and dinosaur remains have been found from Del Puerto Canyon in the north to Orchard Peak, 135 miles to the south. Most of these discoveries were in the Late Cretaceous Moreno Formation.

FIGURE 5.37
The Panoche Hills as seen
from a commercial jet.
Photo by the author.

Being in the rain shadow of the Coast Range, San Joaquin Valley's hills are extremely hot and dry in the summer. These somewhat grass-covered, rolling, and sometimes steep-sided hills are essentially a desert, and hardly a tree can be found. During dry years the grass may be grazed off, exposing the soil and rock to erosion. During the winter rains the wet clay soil makes it almost impossible to get vehicles into the area, so most fieldwork must be done in the hot, dry season from May to October.

William Brewer was in this area in July 1861 doing fieldwork for the California Geologic Survey. In the following narrative he describes the conditions he encountered while going to and from the mercury mines in the New Idria area:

> Then we struck across the plain of the Panoche. I wish I might describe the ride that you might realize it, but words are tame. The temperature was as high as any traveler has noted it (so far as I know) on the deserts of Africa or Arabia. Hour after hour we plodded along— no tree or bush. A thermometer held in the shade of our own bodies (the only shade to be found) rose to 105 degrees—it was undoubtedly at times 110 degrees, while in the direct rays of the sun it must have fluctuated from 140 to 150 or 160 degrees. I think, from other observations, it must have risen to the last figure!

FIGURE 5.38
The Panoche Hills, March 3, 1940. Photo by Arthur Drescher; courtesy of the Vertebrate Paleontology Department, Natural History Museum of Los Angeles County.

Early Discoveries: 1918 to 1936

The history of discovery in the western San Joaquin Valley begins with the finding of two tail vertebrae from a type of giant marine lizard known as a plotosaur (a type of mosasaur). They were found in Fresno County, in the Moreno Formation of the Panoche Hills, in 1918 or 1920 by Herman G. Walker of Oakland (Camp 1942a).

In the summer of 1935 Neal Johnstone Smith found an ichthyosaur rostrum (beak) fragment in a radiolarian chert cobble near the mouth of Corral Hollow Creek in the northwestern San Joaquin Valley (Camp 1942b). The cobble had apparently washed into the valley from the Coast Range Province to the west, so technically this is a Coast Range fossil. Camp (1942b) named the specimen *Ichthyosaurus franciscus.*

Sometime prior to January 1936 an oil geologist discovered the remains of a badly broken skull that has been tentatively identified as an ichthyosaur in what Chester Stock, a Caltech paleontologist, recognized as Jurassic or Cretaceous rocks in the northwest corner of Kern County (see fig. 5.40). The remains were catalogued, however, as being Triassic in age and from the Hosselkus Limestone. Because there are no such beds in Kern County, Stock was correct: they are surely from Cretaceous Great Valley Group rocks.

FIGURE 5.39 (LEFT)
Typical highly inclined strata (exposed by wave erosion on San Luis Reservoir) of the Great Valley Group in the hills west of the San Joaquin Valley. Photo by the author.

FIGURE 5.40 (BELOW)
Fragment of ichthyosaur(?) skull found in Kern County in the early 1930s. Courtesy of the Vertebrate Paleontology Department, Natural History Museum of Los Angeles County; photo by the author.

Bennison: 1936 and 1937

On June 11, 1936, Allan Bennison, then a high school senior, found the first dinosaur in California in the hills west of Gustine. Associated with the remains were three mosasaur vertebrae (*Kolposaurus* [now *Plotosaurus*] *tuckeri*), discovered by Bennison's high school science teacher, M. Merrill Thompson. Thompson alerted the paleontology department at UC Berkeley of Bennison's find. Berkeley paleontologist Samuel P. Welles, curatorial assistant Curtis Hesse, and artist Owen J. Poe assisted in the excavation of what proved to be hadrosaurian remains consisting of several vertebrae (see fig. 5.43). In addition, the rancher who owned the land, John Hammond, was using a femur (thigh bone) as a doorstop. The bones went to the UCMP, although Thompson retained a couple of the vertebrae (now assumed lost). Hesse and Welles (1936) announced the discovery in print (see fig. 1).

ALLAN P. BENNISON (1918–)

FIGURE 5.41
Allan Bennison in 2000
at the site of his 1936
hadrosaur discovery.
Photo by the author.

Allan Bennison's love of fossils stems from his high school days in Monterey, when he went on collecting trips in the area. In the middle of his high school years his family moved to Gustine, in the northern San Joaquin Valley. From 1934 to 1940 he rode his bicycle, often seventy miles or more round-trip, to the hills on the west side of the valley in search of fossil shells. While hiking in these hills on June 11, 1936, at the age of seventeen, he found the first dinosaur remains in California. After scientific study they were determined to be from a hadrosaur (a duck-billed dinosaur).

In 1937 Bennison made another important find, this time from a marine reptile: the best-preserved mosasaur skull ever found in California. It was later published and named for him in C. L. Camp's *California Mosasaurs* (1942a), with the

In 1937 Bennison scored again. This time he discovered the best-preserved mosasaur skull ever found in California. Thompson started excavation, followed by Hesse, Welles, Poe, and Bennison. The skull was beautifully prepared by Martin Calkin and later published in C. L. Camp's *California Mosasaurs* (1942a), illustrated by Poe. It was named *Kolposaurus* (now *Plotosaurus*) *bennisoni*.

FIGURE 5.42 (OPPOSITE LEFT)
Paleontologist Samuel Welles helps prepare the mosasaur *Plotosaurus bennisoni* discovered by teenager Allan Bennison in 1937 in the hills west of the San Joaquin Valley. *Oakland Tribune* photo, courtesy of Paul Welles and the University of California Museum of Paleontology.

FIGURE 5.43 (OPPOSITE RIGHT)
Seventeen-year-old Allan Bennison in 1936 at the site of discovery and with bones of California's first dinosaur. Courtesy of the University of California Museum of Paleontology.

FIGURE 5.44 (LEFT)
From left to right: Curtis Hesse, Allan Bennison, and M. Merrill Thompson sit beside a pile of hadrosaur bones, California's first dinosaur remains. Courtesy of the University of California Museum of Paleontology.

name *Kolposaurus* (now *Plotosaurus*) *bennisoni.* Bennison donated much of his extensive collection of fossils found in those early years to the California Academy of Sciences and to the Museum of Paleontology at Berkeley.

In 1940 Bennison graduated cum laude from the University of California at Berkeley with an A.B. degree in paleontology and a minor in geology. His professional career included teaching geology at Antioch College in Ohio and work as a photogrammetrist for the U.S. Geological Survey and as a petroleum geologist in South America. He later helped form a geo-resources company. He has published numerous papers and has extensive experience in geologic mapping. His geologic highway maps published by the American Association of Petroleum Geologists (of which he is an honorary member) are highly regarded. Now in his eighties, he continues his career as a consulting geologist and in his spare time still enjoys searching for fossils. In 2000 he discovered the first Jurassic plesiosaur tooth from California.

Welles, Paiva, and Tucker: 1937

Also in 1937 rancher Frank C. Paiva found California's first plesiosaur on his land in the now famous fossil locale, the Panoche Hills. That spring the remains were brought to the attention of William M. Tucker of Fresno State College. (According to a friend of Tucker's, Theodocia Wilmoth McKensie [pers. comm. 2000], up to the age of thirty Tucker had been

FIGURE 5.45
William M. Tucker (wearing
pith helmet) of Fresno State
College heads up the first
excavation crew in the
Panoche Hills, July 1937.
Photo by Chester Stock;
courtesy of the Vertebrate
Paleontology Department,
Natural History Museum
of Los Angeles County.

a farmer in Indiana. Apparently some of his relatives jokingly commented that farming was all he had the brains to do. He promptly enrolled in college and in a few years graduated with a doctoral degree.) After partial excavation by Tucker, Welles and his Berkeley crew were called out for assistance. According to Welles (1943), "At the time of discovery, only a few posterior cervical [neck] vertebrae were exposed at the very bottom of a V-shaped gully. The skeleton lay upon its right side along a bedding plane. The head and neck were upslope and covered by only a foot or so of matrix, but the body and tail extended 15 feet under the eastern bank." They apparently had to dig a tunnel to get the entire specimen out. There was evidence that some of the bones had been displaced by scavengers, and "the teeth of sharks and shells of ammonites were collected a few feet from the skeleton, and these animals may have torn away bits of flesh and bone from the carcass."

The following is based on an account of the dig from the notes of Samuel P. Welles: The team left for the Panoche Hills in the heat of summer on August 8, 1937. Their ten-year-old Cadillac broke down in Gilroy, but with a new distributor cap they were soon on their way again. At the site, Welles was aided by rancher Frank Paiva, assistant Lloyd Conley, workers Charley Fowler and Joseph Michael, and cook E. Stapp. The next day was spent exposing the right hind flipper. On the following day they worked until 11:30 P.M. getting plaster jackets on the flipper, the forearm, and part of the vertebral column. On the twelfth they enlisted the help of a mule and scraper (see fig. 5.47), then they used black powder (an explosive) to blast away surrounding rock. They found another flipper.

FIGURE 5.46 (ABOVE)
William M. Tucker, E. W. Moore, and Frank C. Paiva at plesiosaur site in July 1937. Courtesy of the Vertebrate Paleontology Department, Natural History Museum of Los Angeles County.

FIGURE 5.47 (LEFT)
Using a mule and scraper to smooth out a path so that the jacketed specimens can be removed. Photo by Chester Stock, July 1937; courtesy of the Vertebrate Paleontology Department, Natural History Museum of Los Angeles County.

Welles comments: "Weather is exceedingly hot & dry. Hottest I've ever experienced, the quarry is the coolest, or rather least hot, spot around. Sun comes up at 5:20 & [it] goes to work at once [on us]. . . . Al & I went back to the quarry & got 2 more shots [blasts] in before the mule arrived. Dr. & Mrs. Stock visited us at 3 or so, took about 1 doz. pictures and ran off the road. We helped them out and received a parting gift of peaches." (Chester Stock had left UC Berkeley in 1926 and was teaching at Caltech in Pasadena. In 1939, without informing his colleagues at Berkeley, Stock organized new exploration and digs in the Panoche Hills. He later apologized to his Berkeley colleagues, saying he felt bad about going into "Berkeley field territory.")

On August 14 UCMP paleontologist Charles L. Camp and his wife, Harriet, arrived at the dig. They helped work out some of the ribs and the left hind flipper. Camp found several small

rounded stomach stones (gastroliths) as well. The next day they looked at a mosasaur that Paiva and Tucker had found and which was later named *Kolposaurus* (now *Plotosaurus*) *tuckeri*.

On the seventeenth they were back working on the plesiosaur. They constructed two tunnels under the block containing the tail and in the evening made a plaster jacket around it. The following morning they put two braces on the tail block and then worked on the left hind flipper. Welles commented that "yesterday Frank and Lloyd had trouble loading the mule—she put her foot thru the rear cab celluloid [an early plastic] and caused quite a rumpus. Last night (5:30) she broke out and we had to go round her up down on the flat in the Cad [Cadillac]." Later that day they jacketed both rear flippers and put a lifting brace on the tail block. They were in a race with time, as the bank containing the left hind flipper was drying out and threatening to cave in.

On Friday the twentieth they tunneled under that flipper and turned the block containing the right hind flipper. "The LH [left hind flipper] was more troublesome," Welles reported, "as our tunnel was too deep & we tried to break too much earth. Bailing wire held until matrix was chiseled away & the block turned." Paiva and Conley completed building the sled they planned to use to help slide the jackets containing the bones up the steep gully to the car.

CLOCKWISE FROM
OPPOSITE PAGE

FIGURE 5.48
Charles L. Camp in the field
in 1942. Photo courtesy and
from the collection of Kevin
Padian, University of
California Museum of
Paleontology.

FIGURE 5.49
Lloyd Conley struggles to
load a block on a sled so it
can be moved to the vehi-
cles. Courtesy of the
University of California
Museum of Paleontology.

FIGURE 5.50
Each part of the specimen
had to come out in a separate block. Here Sam Welles
and Lloyd Conley under-
mine one of the blocks so
that it can be encased in
burlap and plaster for
removal. Courtesy of the
University of California
Museum of Paleontology.

FIGURE 5.51
Frank Paiva, Lloyd Conley,
Jean Johnson, and Harriet
Welles inspect one of the
jacketed specimens ready for
transport. Courtesy of the
University of California
Museum of Paleontology.

FIGURE 5.52
Jean Johnson, Irmgard Johnson, Harriet Welles, and Lloyd Conley pull and push the loaded sled up the steep gully to vehicles above. Courtesy of the University of California Museum of Paleontology.

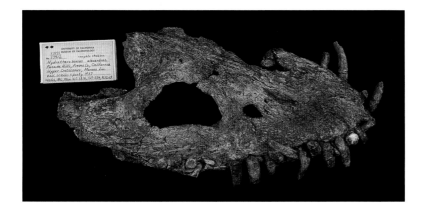

FIGURE 5.53 (ABOVE)

Sam Welles considered the excavation of *Hydrotherosaurus alexandrae* (which he later published as his Ph.D. thesis) to be his most challenging dig. From Welles 1943.

FIGURE 5.54 (LEFT)

The skull of *Hydrotherosaurus alexandrae.* Courtesy of the University of California Museum of Paleontology; photo by the author.

On the twenty-fifth they slid up one block and then spent the rest of the day getting two tunnels under the block containing a front flipper. Still working on the twenty-ninth, Welles comments: "The upper 2 × 6 [board] worked under the block very nicely but the groove for the lower [board] cut into a slick bedding plane & the whole block slid down the bedding slope on this new plane, banging the pick handle against my leg. Most fortunately no one was trying to work under the block at the time!" It took the entire crew and three jacks to gradually work this large block up the hill (see fig. 5.52).

The next day Welles complained: "The damn 'casting plaster' didn't set so I suppose the program is to wash it all off. —I wrapped while Frank and Lloyd took off all the old lime. It had a very sharp odor & was quite caustic on sore hands." (It sounds like they were using pure lime, instead of casting plaster.) After lunch they used the tractor to help remove a large jacketed specimen. Although the tractor seemed to have plenty of power, it didn't have enough traction, so Conley helped with a pry bar almost every inch of the way while Welles cleared

SAMUEL P. WELLES (1909–1997)

I had the good fortune of meeting Sam Welles, a scientist at the University of California Museum of Paleontology and one of the old-timers in the field of vertebrate paleontology, before starting this book. My friend Pat Antuzzi had found reptile remains in the Chico Formation, and we brought them to Welles thinking they might be part of a mosasaur skull. Welles immediately recognized them as just that and brought out a cast of Allan Bennison's skull of *Plotosaurus bennisoni* to compare against the fragments. Welles died shortly after our meeting, unfortunately, but in a sense I feel I have continued to get to know him from the writings and field notes he left behind.

Somewhere between 1910 and 1912, when Welles was a little boy, his family moved to Berkeley, where he remained for the rest of his life. He was hired by paleontologist William D. Matthew of the University of California Museum of Paleontology as a preparator

FIGURE 5.55

Sam Welles sitting under the mounted skeleton of the *Hydrotherosaurus alexandrae* that he excavated from the Panoche Hills and later named for University of California Museum of Paleontology patron Annie Alexander. Courtesy of the University of California Museum of Paleontology.

away the banks. They reached the top by 7 P.M. and used the tractor and a ramp made of two-by-twelves to load it on the truck. Welles comments: "Believe me I had a few nervous moments as the truck with all our loot aboard came over these crazy slopes."

After loading the remaining blocks on the truck, they tunneled into the steep bank following the gracefully curving neck vertebrae toward the skull. After discovering the somewhat flattened skull, they spent September 1 exposing more of it and the neck. The next day they plastered the skull and two adjoining blocks and sledded them and the remaining blocks up the hill. On Friday the third they broke camp.

According to Robert Long (pers. comm. 1998), Welles considered the excavation of *Hydrotherosaurus alexandrae,* which he named for Annie Alexander, to be his most challenging dig.

Stock, Drescher, and Crew: 1939 to 1940

For a year and a half no work went on in the hills to the west of the San Joaquin Valley. Then Caltech professor Chester Stock decided to go to the Panoche Hills to look for fossil reptiles. Arthur Drescher was chosen by Stock to be in charge of the field crews.

right after graduation from UC Berkeley, and by the mid-1930s he was leading expeditions into the field. In 1936 Welles's crew excavated the dinosaur discovered by Allan Bennison, the first one in California. The next year he was contacted by William M. Tucker of Fresno State College about a large plesiosaur that had been found in the Panoche Hills and partially excavated by Tucker. Welles, along with Professor Charles Camp of the Museum of Paleontology, led crews to excavate the monster. Because the head and neck went straight down into rock, they had to dig a tunnel to get them out. He named the specimen *Hydrotherosaurus alexandrae* for museum benefactor Annie Alexander. She provided a travel grant to Welles to study the fossil plesiosaurs found in Europe, which helped him better understand the California plesiosaurs (Welles 1952). He later named a mosasaur *Plotosaurus tuckeri* after Tucker.

Welles received his Ph.D. from UC Berkeley in 1940 and became the assistant curator of amphibians and reptiles at the Museum of Paleontology. He taught courses in general and vertebrate paleontology until 1963, continuing his teaching in the field after that. Welles's field trips, which he led from the late 1940s to the mid-1980s, were always popular with the students.

Unlike many of the scientists in this book, Sam Welles specialized in fossil reptiles, excavating ichthyosaurs in Nevada, Triassic vertebrates in the Colorado Plateau, and dinosaurs in Arizona. He discovered the dinosaur *Dilophosaurus.* Welles was working on a paper on a Jurassic plesiosaur discovered in the foothills of the Sierra when he died.

According to Drescher (pers. comm. 1999),

> The field members varied greatly. They consisted of "permanent" people who during their tenure were paid and were more or less on call at any time in 1939, 1940, and the first half of 1941. These were myself, Bob Leard, Betty Smith and Otis Fenner. Betty and Otis were not from Caltech. Then there were Caltech graduate students and undergraduates, [who] were available a few weekends and breaks during the school year and/or a few weeks during the summers. These were Bob Hoy, Bob Wallace, Lloyd Lewis, Joe Rominger, Jack Dougherty, Clyde Wahrhaftig, Tom Fox, Bob Wilmoth, Charles McDougall and Dale Turner. Caltech graduate students Paul Henshaw, Dick Jahns, Bob Wilson, Richard Hopper and John Cushing visited the digs for a day on one occasion.

Frank Skalecky also participated, though Drescher did not recall his presence on the team. Norvel Roper was employed as a photographer during the excavation of the *Trachodon* (*Saurolophus*).

Drescher would assign specific crew members to painstakingly scour designated drainages and ridges, making sure the area was covered thoroughly. He obtained a dog from a local shepherd who lived on Panoche Creek, naming it Temy—short for a fossil dog called *Temnocyon* (Drescher, pers. comm., 2001). The dog became a close companion to all the members of the

FIGURE 5.56
Arthur Drescher (left) and
Otis Fenner stroll through
Fresno on one of their trips
to town for supplies. Photo
from the collection of
Arthur Drescher.

group. The following is Art Drescher's summary (pers. comm. 1999) of the site and work in the Panoche Hills:

> In those days the planted [agricultural] area extended only half of the thirty miles from Mendota to the sheep ranch at Panoche Creek and the only north-south road from Taft to Tracy was Highway 33, "The Westside Highway." It was two-lane and not all of it was paved. The only road other than a few ranch roads west of Highway 33 was a dirt road from Dos Palos southward to where it turned and headed westward through the hills to Mercy Hot Springs. From there on south to the entrance to Panoche Valley there was only a sheep-ranch road. To get to our camp and the fossil beds from either Mendota or Dos Palos we had to go all the way to the sheep ranch and then go back northward closer to the hills on an untracked course.
>
> Initially Mendota was our supply point but the water was unfit for drinking and not even good for plastering and we chose other sources such as Firebaugh, Dos Palos and Los Banos. Fresno became our main Saturday shopping point for food and supplies. Mendota was our last stop on the way to camp and we would stop at the bar for a beer and to listen to the Beer Barrel Polka. Dos Palos became our midweek supply point for water, plaster, lumber et al. On one return trip one of the crew was scalded pretty badly by hot water from the radiator of the truck and he had to be returned to Dos Palos for medical treatment. An irrigation canal crossed the road a few miles south of town and we usually

FIGURE 5.57
The results of a windstorm that blew tents down at the "Gros Ventre" camp in the Panoche Hills, September 9, 1939. Courtesy of the Vertebrate Paleontology Department, Natural History Museum of Los Angeles County.

FIGURE 5.58
Dale Turner (left) and Robert Leard using a horse and scraper to remove over-burden at a mosasaur site, June 1939. Photo by and from the collection of Arthur Drescher.

stopped for a swim to get some relief from the heat, including Betty at some distance from the rest.

There were two other accidents of note at camp. The first was a windstorm which flattened the tents, blew things away and messed things up generally. The second was a fire from the cooking stove that destroyed the main tent.

Our working practice in the summer usually was to start early and spend the morning prospecting, then after lunch to work on specimens under the shade of tarps. In winter in wet spells it was difficult to impossible to get around because of the clayey slippery soil. On one occasion Bob Leard and I were unable to get back over a hump in the ridge we were on and decided to make camp to wait out the rain. During the night a large flock of noisy sheep moved in, surrounding us. We stayed there for three days in the tent with nothing to do except to read and listen to the radio.

ARTHUR DRESCHER (1911–)

Arthur Drescher's leadership and fossil hunting skills have led to some of the most important fossil reptile finds in California. Before coming to the California Institute of Technology, Drescher collected fossils for three summers in South Dakota. He had also worked as a fossil preparator for three years for the South Dakota School of Mines Museum. He received his master's degree in 1939 from Caltech. Chester Stock then chose Drescher to head up his field crew in search of Mesozoic reptiles in the western San Joaquin Valley. From 1939 to 1941, as a result, he spent considerable time in the field.

Arthur Drescher and Robert Wallace discovered the plesiosaur *Morenosaurus stocki,* and Samuel Welles named another plesiosaur, *Fresnosaurus drescheri,* in honor of Drescher's hard

FIGURE 5.59
Arthur Drescher in 1939 excavating the skull of the mosasaur
Plesiotylosaurus crassidens he found in the Panoche Hills. Photo by
Robert Wallace; courtesy of the Vertebrate Paleontology Department,
Natural History Museum of Los Angeles County.

Stock's crew started work early in 1939. Although we have none of his notes, we do have photographs, publications, and some letters. Stock writes that Robert T. White of the Barnsdall Oil Company of Bakersfield had found the remains of a plesiosaur in the Panoche Hills and brought it to the attention of Stock. On February 26, 1939, Stock was in the Panoche Hills taking pictures of his crew—which consisted of Drescher and Lewis—exposing the flipper of the plesiosaur. This specimen is catalogued as the plesiosaur *Aphrosaurus furlongi,* named for Stock's preparator Eustace L. Furlong.

By April Richard (Dick) Jahns, Richard Hopper, Paul C. Henshaw, Jack F. Dougherty, and Robert B. Hoy had joined the crew, and in May Robert Wallace joined as well. Wallace (1999) comments:

> He [Stock] had access to some funds from the Carnegie Institution of Washington, which could be used in field work. It was the only such money in town. So when any of us wanted a weekend out in the hills, Stock could pay for food and our gasoline, and we could use the department car. . . . We were looking for fossils of . . . sea-serpent-type critters. . . . I think it was Art Drescher who made the first find—the nose of a Mosasaur sticking out of the ground. We hiked back over the hills to the truck to get shovels and equipment. Then

work for Caltech. He and Robert Leard also found a hadrosaurian dinosaur, *Saurolophus,* and Drescher discovered the mosasaur *Plesiotylosaurus crassidens.* Besides making important finds, he was responsible for ensuring that all the excavations were conducted properly. Although his crews were relatively small, three to five people at a time, ultimately he was responsible for at least nineteen people. Overseeing camping, cooking, vehicle maintenance, fossil hunting, excavation, and just plain trying to stay alive in a waterless desert was a huge responsibility.

After completing his minor thesis in paleontology in 1941 he got a job with the U.S. Geological Survey in San Luis Obispo. He later took a geology job in Texas. Drescher's career included work as an exploration geologist in the Southwest for the Bear Creek Mining Company (Kennecot); as a petroleum geologist for Shell Oil Company out of Bakersfield; as a district geologist in Laramie, Wyoming, for Union Pacific Railroad Company; and seventeen years with the Lone Star Steel Company in Texas as division superintendent for ore and coal. He went on to the Homestake Mining Company as chief exploration geologist in North America and Australia. He retired in 1978 and presently lives in Oregon.

FIGURE 5.60
From left to right: Arthur Drescher, Richard Jahns, Jack F. Dougherty, Paul C. Henshaw, Richard Hopper, and Robert Hoy excavate the plesiosaur *Aphrosaurus furlongi* in April 1939. Photo by Jack F. Dougherty; courtesy of the Vertebrate Paleontology Department, Natural History Museum of Los Angeles County.

while on the way back to the Mosasaur I saw three vertebrae (neck bones) of a Plesiosaur sticking out of the ground. . . . Here we had two major fossil finds of big creatures in that one weekend. What Excitement! . . . we dug very carefully . . . finally got down with a toothbrush and small tools to clear details.

We put burlap coated with plaster-of-Paris over them to protect them, and ended up leaving both of them at the site for many months. . . . [In] the summertime . . . we had big cloth canopies over the site to protect the crew from the San Joaquin Valley sunshine.

It became a real "dig." . . . The skeleton was complete except for the head. . . . The head was gone, but all the rest was preserved . . . including the stomach material with bones of little critters the Plesiosaur had eaten.

FIGURE 5.61 (ABOVE LEFT)
Morenosaurus stocki on display today at the
Natural History Museum of Los Angeles
County. Courtesy of the Vertebrate Paleon-
tology Department, Natural History Museum
of Los Angeles County; photo by the author.

FIGURE 5.62 (ABOVE RIGHT)
Three neck vertebrae that led to a nearly
complete plesiosaur found by Robert Wallace
and Arthur Drescher on May 22, 1939. This
plesiosaur was later named *Morenosaurus
stocki* for Chester Stock. Photo by Robert
Wallace; from the collection of Arthur
Drescher.

FIGURE 5.63
From left to right: Arthur Drescher, Robert Hoy, and Robert Wallace
work on the neck of the plesiosaur, which is angling into the earth.
Photo by Robert Wallace; courtesy of the Vertebrate Paleontology
Department, Natural History Museum of Los Angeles County.

The plesiosaur became the mounted, full specimen that is on display today at the Natural
History Museum of Los Angeles County (see fig. 5.61). Drescher's mosasaur was named *Plesiotylosaurus crassidens* (see figs. 5.64 and 5.65).

On an earlier trip to Oregon, Stock had had what appeared to be a serious gall bladder attack, after which he left most of the fieldwork to others. In June Robert M. Leard came on

FIGURE 5.64 (LEFT)
From left to right: Robert Hoy, Arthur Drescher, and Robert Wallace after having excavated the skull of the mosasaur *Plesiotylosaurus crassidens*. Photo by Robert Wallace; from the collection of Arthur Drescher.

FIGURE 5.65 (BELOW)
By May 25, 1939, the plaster jacket has been removed from the mosasaur *Plesiotylosaurus crassidens* and Arthur Drescher is preparing it in the Caltech lab. Photo by Robert Wallace; from the collection of Arthur Drescher.

as Drescher's assistant, and they were joined by Dale Turner, Joe Rominger, Betty Jean Smith, Otis Fenner, and Clyde Wahrhaftig—and let's not forget "Queenie," the mule. Lewis, Jahns, Dougherty, Henshaw, Hopper, and Hoy appear to have left at this time. Mid-June finds the team camped above a steep gully at "Reptile Ridge Camp," complete with canvas tents, a stall for the mule, the Caltech panel truck, and a woody (station wagon). On June 16, 1939, Drescher wrote Stock:

FIGURE 5.66
Overview of camp, June
1939. Photo by Arthur
Drescher; courtesy of the
Vertebrate Paleontology
Department, Natural
History Museum of Los
Angeles County.

FIGURE 5.67
From left to right: Arthur
Drescher, Robert Leard,
Dale Turner, Joe Rominger,
Betty Smith, and Clyde
Wahrhaftig at Reptile Ridge
Camp, June 1939. Photo by
Arthur Drescher; courtesy of
the Vertebrate Paleontology
Department, Natural
History Museum of Los
Angeles County.

Everything is going fine. We are going to Kingsburg today for a pair of jackasses and a scraper. In view of the expense entailed in this matter—$3.00 a day for all equipment— it will be necessary for us to have another advance very soon. . . . On the first day out here Miss Smith [Betty Jean] found some large bones which, on excavation, have turned out to be those of a dinosaur, probably *Trachodon* [*Saurolophus*]. They represent the hind legs, pelvic region, ribs, sacrum, etc. and are still going in. It will require a large excavation because of the steep dip. I hope you will come to see it on your way north. If it turns out as it seems to promise, it will be the best thing we have [found] yet.

As of early July the crew seems to have consisted of Drescher, Leard, Smith, Fenner, Rominger, Turner, and Wahrhaftig. John Cushing joined the team for a short time as well. Wallace, who probably left in June, was photographing the unloading of specimens at Caltech in July (see fig. 5.81). Drescher had written Stock on June 30:

FIGURE 5.68
"Queenie" the mule trans-
ports four-hundred-pound
jacketed hadrosaur remains
on scraper/sled. Robert
Leard is guiding the specimen, while Dale Turner
handles the reins. Photo by
and from the collection of
Arthur Drescher.

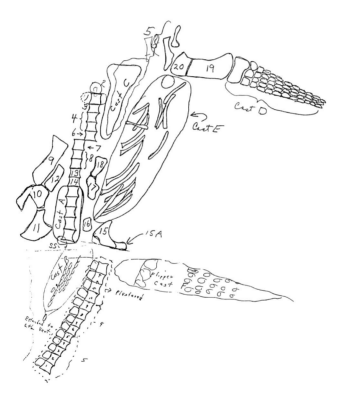

FIGURE 5.69
Half-sized field sketch of
exposed plesiosaur made
by Otis Fenner. Courtesy
of the Los Angeles County
Museum of Natural History.

This week we have followed the practice of prospecting in the mornings and working on
the plesiosaur in the afternoons. Our prospecting has yielded a part of another skeleton,
possibly a Mosasaur, though that is just a guess. Miss Smith found it. . . . [If this is site 331,
identified in Stock 1939, it had fish remains in its stomach.] Wahrhaftig found a small
string of vertebrae of a Plesiosaur. . . . At the main Plesiosaur, further work has disclosed

CHESTER STOCK (1892–1950)

It is interesting that in George Gaylord Simpson's "Biographical Memoir of Chester Stock," Chester Stock's important work on the study of Mesozoic reptiles in California is not so much as mentioned. He did so much good work on Cenozoic mammals that his contributions in the excavation, curation, scientific study, and publication of fossil reptiles were all but forgotten.

Chester Stock was born to a struggling German family in San Francisco. His hair gained him the nickname "Red," and as a youth he operated a newspaper stand at the corner of California and Battery Streets and played the tuba in a local band. Before the California Academy of Sciences burned—along with the Stocks' house—during the earthquake and fire of 1906, Chester may have gotten his inspiration for vertebrate paleontology from the painted models of the mammoth, plesiosaur, and ichthyosaur exhibited there.

During the catastrophe of earthquake and fire, Stock, at the age of fourteen, was mustered for a short time into the California National Guard. Shortly afterward, partly because of family hardships caused by the loss of their home, Stock quit school to go to work in the Union Iron Works. Hard work and a case of malaria wreaked havoc on his health, so in 1910 he went back to high school and soon graduated with honors. That fall he enrolled in the University of California at Berkeley to study medicine, but, probably owing to the influence of paleontologist John C. Merriam, Stock changed his major to paleontology.

He received his B.S. degree in 1914 and his Ph.D., also from UC Berkeley, in 1917. In 1918 he published his dissertation on the Hawver Cave Pleistocene mammals found near Auburn. Fossil mammals were to become the focus of his expertise and devotion in paleontology for the rest of his life. Most of his work with the University of California, Caltech, Natural History Museum of Los Angeles County, U.S. Geological Survey, and at Rancho La Brea was in mammalian paleontology. His work on Mesozoic reptiles in California was in a sense a diversion, but definitely an important one.

Merriam recognized that Stock was an upcoming star in vertebrate paleontology. A letter dated March 2, 1914, to Professor Bennitt M. Allen at the University of Kansas exemplifies Merriam's early confidence in his student: "Stock has definitely committed himself to a program of scientific work in paleontology and having an excellent equipment of personal qualifications, industry, and definiteness of purpose, I expect to

FIGURE 5.70
Chester Stock in 1944. Charcoal on paper by artist Peter Van Valkenburgh, from the collection of the author. Photo by the author.

a complete left limb, including the flipper . . . the work has not progressed enough to disclose the skull, or to prove its absence. It is a beautiful specimen. . . . Wahrhaftig has talked to his folks and tells me that they did not give out any information about our work. In fact it seems that when Mrs. Wahrhaftig [Wahrhaftig's mother] was talking to Dr. Tucker, he already knew about us. He must have received his information some other way. I guess that no harm has been done, since we have not seen any of Camp's parties yet.

see him one of the best men in the country in a few years." Stock stayed on at Berkeley as an assistant until 1919, when he became an instructor.

In 1920 Merriam left for Washington, D.C., to become president of the Carnegie Institute. Assistant Professor Stock remained at Berkeley as Merriam's successor until 1926, when he accepted a full professorship at the California Institute of Technology. Here Stock became a research associate of the Carnegie Institute, and through the Carnegie Merriam helped finance much of Stock's work. According to Simpson, "Merriam's influence on Stock was profound . . . [and they had] an odd sort of intellectual symbiosis."

In the years 1939 to 1944 Stock spent much of his energy on fossil reptiles rather than mammals, sparked, perhaps, by a 1937 visit to Samuel Welles's field camp. Mosasaurs, plesiosaurs, and even a dinosaur had already been discovered in the hills on the west side of the San Joaquin Valley, but Stock thought he could find more—and with the help of Caltech and Carnegie funds, he could finance the crews to do so. The first two years were extremely productive, and although fieldwork tapered off after 1940, the jobs of publishing and making museum mounts of some of the specimens continued until 1944.

Stock's crews excavated the most complete hadrosaurs in California, a complete skeleton of a mosasaur, and a nearly complete plesiosaur, all of which were good enough for full mounts in the museum at Caltech. The mosasaur and the plesiosaur can be seen today at the Natural History Museum of Los Angeles County. All together, Stock's crews found the remains of two hadrosaurs, a turtle, ten plesiosaurs, and a dozen mosasaurs. Stock was honored when Samuel P. Welles of the Museum of Paleontology at Berkeley named the plesiosaur *Morenosaurus stocki* after him.

As a teacher and a person, Stock was well liked, even loved. His student Paul C. Henshaw described his ability to "spellbind a class of three or four students or an audience of hundreds with equal ease." G. H. Curtis said of his lectures, "He was beautifully organized and had a wonderful delivery." Art Drescher remarked, "I can't praise him too highly. He rates in the top three or four outstanding people I have known, not just personally but as a kind, considerate, helpful, friendly, cheerful, just plain nice person."

Stock received many accolades in his life. He served as president of several professional organizations, including the Paleontological Society (1945), the Society of Vertebrate Paleontology (1947), and the Geological Society of North America (1950). "Stock," said his colleague Lloyd C. Pray, "was the Pied Piper of vertebrate paleontology. When he talked about vertebrate paleontology you knew it was from a man who not only knew the subject but was experienced in it."

Camp was a Berkeley paleontologist. Evidently the entire Stock crew knew they were in "Berkeley territory."

Once a specimen was found, the work had only just begun, for it then had to be exposed to learn the extent and completeness of it and then be prepared for transport. The excavation, jacketing, transport, and preparation for display of the mosasaur *Plotosaurus tuckeri* was documented in photographs. These, along with Art Drescher's narrative (pers. comm. 1999), tell a story that is typical for the retrieval of large fossil skeletal specimens, both at that time and, to a great degree, even today.

FIGURE 5.71
From left to right: Robert Hoy, Arthur Drescher, and Robert Wallace dig back into the hillside to expose a skeleton, May 1939. Photo by Robert Wallace; courtesy of the Vertebrate Paleontology Department, Natural History Museum of Los Angeles County.

FIGURE 5.72
Arthur Drescher carefully brushes away loose sediment from vertebral area. Photo by Robert Leard; courtesy of the Vertebrate Paleontology Department, Natural History Museum of Los Angeles County.

In the Panoche Hills the strata were dipping 10 or 20 degrees. The first step in collecting was to follow the bones down dip a short distance in order to judge how the specimen was projecting ahead. With this knowledge the next step was to remove as much overburden as possible down to a safe distance (several inches or more) above the bone layer. This was a heavy pick and shovel job but sometimes required using a horse and scraper or even black powder explosives in hand drilled holes. From there on down we used light weight Marsh picks to about an inch or so above the specimen and then screw drivers, small chisels, awls and whisk brooms and paint brushes. The main tool was a screw driver, the metal shaft of which had been given two right angle bends in a welding shop. This was a great knuckle saver.

FIGURE 5.73
Betty Smith inspects fully exposed, pillared specimen ready for plaster and burlap jacket, June 1939. Photo by Robert Leard; courtesy of the Vertebrate Paleontology Department, Natural History Museum of Los Angeles County.

FIGURE 5.74
From left to right: Arthur Drescher, Dale Turner, Betty Smith, and Otis Fenner drill holes for bolts to hold the fossil to the wood framework, June 1939. Photo by Robert Leard; courtesy of the Vertebrate Paleontology Department, Natural History Museum of Los Angeles County.

FIGURE 5.75
Arthur Drescher and Betty Smith in the process of jacketing and drilling more holes. Photo from the collection of Robert Wallace.

FIGURE 5.76

From left to right: Otis Fenner, Dale Turner, Clyde Wahrhaftig, and Betty Smith bolt the top of the wooden framework to the bottom to help hold the jacket. Photo by and from the collection of Arthur Drescher.

FIGURE 5.77

Shaded quarry with undercut, jacketed specimen and tripods used to lift specimen, July 1939. Courtesy of the Vertebrate Paleontology Department, Natural History Museum of Los Angeles County.

FIGURE 5.78

Joe Rominger, Robert Leard, Clyde Wahrhaftig, and Betty Smith use the tripod and block and tackle to turn the four-thousand-pound specimen. Photo by Arthur Drescher; courtesy of the Vertebrate Paleontology Department, Natural History Museum of Los Angeles County.

FIGURE 5.79

In July 1939 Arthur Drescher wrote: "We have built a road up to the top of the hill above the quarry. Since the package is so large—ten feet by four feet—and since it will weigh almost three thousand pounds, it will not be possible to carry it in either of the field cars." Photo by Robert Leard; courtesy of the Vertebrate Paleontology Department, Natural History Museum of Los Angeles County.

FIGURE 5.80

Using all means necessary to bring the sled containing the four-thousand-pound speci- men to the truck, July 1939. Photo by Robert Leard; courtesy of the Vertebrate Paleontology Department, Natural History Museum of Los Angeles County.

FIGURE 5.81

From left to right: Bob Wilson, E. W. Moore, Otis Fenner, Robert Leard, and Art (the superintendent of grounds at Caltech, last name unknown) unloading the specimen with block and tackle. Photo by Robert Wallace; from the collection of Arthur Drescher.

Ideally it would be best to include the entire specimen in one package but this was generally impractical in order to have packages of manageable size and weight. The largest single package we handled was a mosasaur [the *Plotosaurus tuckeri* also documented in photographs] that could not be properly divided. It weighed 3780 pounds. In order to treat the bones and to determine how and where to divide the specimen into smaller packages it was necessary to nearly totally expose the specimen. This also allowed the photographing and/or mapping of the specimen which aided the reassembly in the laboratory.

As bones were exposed they would dry out and when thoroughly dry they would be impregnated with thinned shellac applied liberally by gently patting with a soft paint brush. When dry this greatly strengthened them. The first portion of a specimen to be packaged was then excavated so as to leave it on a pedestal. This was undercut an inch or two on all sides a safe distance below the bone layer. The next step was to completely and liberally coat the entire surface with wet toilet paper applied by poking the paper with a wet whisk broom into all the depressions. The specimen is then ready for jacketing. Six inch wide strips of burlap gunny sacks soaked in a fairly thick plaster of Paris slurry are laid in an overlapping fashion on the surface and as much as possible into the undercut, poking it with the fingers into all depressions which when hardened formed a tight strong jacket. Sometimes the package was further strengthened by the inclusion of bracing strips in the plaster jacket. Turning the package upside down so as to work on the under surface was a tricky procedure which varied from case to case and required a lot of innovation and prayer, involving tunneling, drilling bolt holes through the block, making clamps using two by fours and bolts, and use of tripods, cables, winches, block and tackle, and muscle. Once turned over additional matrix was removed to reduce the weight as much as feasible. The jacketing process was then repeated on the under surface.

Getting the packages to a loading point was a laborious job. Some could be hand-carried, but usually they were dragged on a skid using manpower, horsepower or machine power. Sometimes a skidway had to be built for short steep pitches.

The group had been having trouble with the panel truck that carried their camping gear and food supplies: on one occasion it lost all forward gears and they were forced to drive in reverse for over twenty miles (Robert Hoy, pers. comm. 1998). Leard, in a letter to Stock of July 11, reported that they had collected the poorly preserved skull of a mosasaur and further remarked: "The truck is in fair shape, but all five of the tires are worn smooth, one shows fabric, and should be replaced as soon as possible. We have had 2 flat tires in the past 4 days. We purchased 5 gal. of oil as the truck uses about one qt every 65 miles and we could save 5 cents/qt by buying it in bulk." In a letter written that same day, Drescher, discussing the preparations to transport the mosasaur to Caltech, mentioned that "I have . . . written to Mr. Furlong to try to arrange to use the large Institute stake truck."

On July 26, Leard wrote Furlong: "Again our funds are rather low and we should have

FIGURE 5.82

Preparator William Otto readies the mosasaur for display at the California Institute of Technology. Courtesy of the Vertebrate Paleontology Department, Natural History Museum of Los Angeles County.

FIGURE 5.83

William Otto and Chester Stock (below) inspect a mounted mosasaur at the California Institute of Technology. This species was named *Kolposaurus* (now *Plotosaurus*) *tuckeri* by Camp (1942a). Courtesy of the Vertebrate Paleontology Department, Natural History Museum of Los Angeles County.

another check as soon as possible. Today I sent the $75 check made out to Mr. Drescher to him at C.I.T. [Caltech] and when he gets it he will send it to me, but that will take some time, and as all members of the party, including myself are 'flat broke' we will be in difficulty when our present funds are exhausted which will be on July 30–31."

In the heat of August Frank Shalecky joined the crew, which then consisted of Leard, Fenner, Rominger, Turner, and Wahrhaftig. Leard left the field on August 8, after Drescher returned from a short absence.

August 13 found them digging out a plesiosaur. Photographs show that the animal was very near the surface; although this makes excavation easy, it also means that because of weathering

FIGURE 5.84 (RIGHT)
In September 1939 Clyde Wahrhaftig (left) and Otis Fenner, along with Arthur Drescher's dog Temy, excavate a large plesiosaur from a very steep site. Here a flipper is jacketed and the rib section has been prepared for covering. Photo by Arthur Drescher; courtesy of the Vertebrate Paleontology Department, Natural History Museum of Los Angeles County.

FIGURE 5.85 (BELOW)
This photograph, taken February 18, 1940, shows little new grass in the Panoche Hills. Photo by Arthur Drescher; courtesy of the Vertebrate Paleontology Department, Natural History Museum of Los Angeles County.

FIGURE 5.86
On their way in to the ple-
siosaur site on February 11,
1940, Robert Leard repairs
the "road" while Temy looks
on. Photo by Arthur
Drescher; courtesy of the
Vertebrate Paleontology
Department, Natural
History Museum of Los
Angeles County.

the specimen is typically not in good shape. In September they found another plesiosaur on a very steep hillside, requiring the building of a trail and deeper excavation.

By September, the beginning of the fall semester at Caltech, the crew had dwindled to Drescher, Fenner, and Wahrhaftig. In a letter of December 21 Drescher informed Stock: "We collected one specimen—114 vertebrae, including a complete tail."

Photographs show that Drescher, Leard, and Smith were again in the field on February 11, 1940. There was little grass, so evidently rainfall had been scarce. However, they did have to do some road building.

The next day they had already started work on a plesiosaur located on the crest of a steep ridge. A fault had displaced the specimen so that two quarries had to be made. Drescher wrote in a postcard to Stock dated February 20: "we are probably progressing towards the tail of this specimen. We have one fine flipper and girdle"; he further remarked that he needed two more weeks on the specimen's tail to complete the work. This plesiosaur turned out to be *Morenosaurus stocki*.

By March, Smith seems to have departed. In a letter to Stock dated April 23, Drescher wrote: "report about a minor accident that occurred to the young man [B. W. Operholser] whom I picked up at the YMCA in Pasadena. . . . he was badly scalded when he opened the radiator cap of the car when it was hot from climbing up the hill to camp. He received some burns on his forehead that took off some of the skin, and his face was quite swollen." In a rather humorous letter to Drescher in the field dated June 10, Stock wrote: "I have just had

FIGURE 5.87
Site of plesiosaur "quarries" located on steep ridge, February 22, 1940. Photo by Arthur Drescher; courtesy of the Vertebrate Paleontology Department, Natural History Museum of Los Angeles County.

FIGURE 5.88
Robert Leard works at exposing a plesiosaur at the two quarries separated by a fault. Photo by and from the collection of Arthur Drescher.

FIGURE 5.89
Temy stands guard and Betty Smith observes as Robert Leard exposes the plesiosaur. Photo by Arthur Drescher; courtesy of the Vertebrate Paleontology Department, Natural History Museum of Los Angeles County.

FIGURE 5.90
Robert Leard prepares ple-
siosaur pectoral flipper,
February 22, 1940. Photo by
Arthur Drescher; courtesy of
the Vertebrate Paleontology
Department, Natural
History Museum of Los
Angeles County.

FIGURE 5.91
After carefully laying wet
tissue paper over the speci-
men, Robert Leard (left)
and Arthur Drescher wrap it
in burlap and plaster, March
19, 1940. Photo by Arthur
Drescher; courtesy of the
Vertebrate Paleontology
Department, Natural
History Museum of Los
Angeles County.

an inquiry . . . concerning a small accident that has been reported upon the Industrial Com-
mission. It was reported as a scalp wound. The victim was B. W. Operholser. I told the office
that so far as I knew there might still be wild Indians in the coast regions of California, but
I was not aware of the fact that Mr. Operholser was employed by our party."

Records show that Leard found a saurian tail on March 1, 1940. Photographs depict a site
in a gullied badlands area below a very steep cliff. This "saurian" turned out to be the mosasaur
Kolposaurus (now *Plotosaurus*) *tuckeri.*

ROBERT MARSHALL LEARD (1915–1966)

Robert Leard was a prominent player in the discovery of Mesozoic reptilian remains in California. Born in Pittsburgh, Pennsylvania, he was an only child. In 1935 and 1936 he went to Caltech for some of his undergraduate work, but finished at UCLA with a B.A. in geology in 1940. He worked for Caltech professor Chester Stock on his paleontological crews in the Panoche Hills in the late 1930s and early 1940s. When crew chief Arthur Drescher was not present Leard headed up the crew. Leard has to his credit the discovery of eight mosasaur specimens: three *Plesiotylosaurus crassidens,* two *Plotosaurus tuckeri,* and three that remain unidentified. He also discovered the remains of three different plesiosaurs: *Aphrosaurus furlongi, Fresnosaurus drescheri,* and *Morenosaurus stocki.* He helped in the excavation of California's most complete dinosaur, a duck-billed plant eater called *Saurolophus.* He later traveled to Mexico to do more paleontological digs.

FIGURE 5.92
Robert Leard works on plesiosaur *Morenosaurus stocki* at night by the light of two Coleman lanterns. Courtesy of the Vertebrate Paleontology Department, Natural History Museum of Los Angeles County.

FIGURE 5.93
Robert Leard talks to a sheepherder next to Caltech truck above one of their more difficult sites (marked with X in extreme right center). Photo by Arthur Drescher; courtesy of the Vertebrate Paleontology Department, Natural History Museum of Los Angeles County.

According to Frank Q. Newton, a close friend later in Leard's life, Leard could not get into the service in World War II because he was diabetic. He wanted to aid in the war effort, however, so he worked in army ordnance. In 1943 he went to work for the Naval Ordnance Test Station in Pasadena, where he became head of quality assurance and eventually had nearly a hundred people working under him. He worked there until his early death.

Leard was married to Letty (Betty Lee) E. McCaskill. He was an avid collector of revenue stamps and antique firearms as well as an expert rifleman. He was vice president of the American Revenue Association in 1963–1964 and president from 1964 until his death (*American Revenuer*, Oct. 1966, 82). Newton comments that he was a quiet, studious, meticulous, no-nonsense man who smoked a pipe and (unlike his days in the Panoche Hills) sported a crew cut.

Leard's contributions to Mesozoic vertebrate paleontology in California are enormous. Unlike Stock, Furlong, and Drescher, his fellow diggers in the Panoche Hills, Leard never had a fossil reptile named in his honor. It is my hope that one day a new species found in California will bear his name.

FIGURE 5.94
Arthur Drescher pours shellac on the folded skeleton of mosasaur *Kolposaurus* (now *Plotosaurus*) *tuckeri*. Photo by Arthur Drescher; courtesy of the Vertebrate Paleontology Department, Natural History Museum of Los Angeles County.

On April 23 there is a note in Stock's letters of a "dinosaur dig." As summer approached and the dig continued, the heat began to take its toll. A canvas tarp was stretched over the dig to provide some respite from the sun. In May, Drescher and Leard were joined by Norvel Roper, a photographer hired to document the excavation.

Stock visited the Panoche Hills *Saurolophus* dig the first weekend of May. Money problems

FIGURE 5.95
Robert Leard rests at camp after using a truck and a wooden sled to drag one of the jacketed specimens of the plesiosaur *Morenosaurus stocki* from the quarry to the top of the hill west of camp. Another jacketed specimen can be seen near the tent. Photo by Arthur Drescher; courtesy of the Vertebrate Paleontology Department, Natural History Museum of Los Angeles County.

FIGURE 5.96
On April 26, 1940, the crew uses black powder (an explosive) in the dinosaur quarry, on a hadrosaur now called *Saurolophus*. Photo by Robert Leard; courtesy of the Vertebrate Paleontology Department, Natural History Museum of Los Angeles County.

continued to plague the crew. In a telegram to Stock sent May 21, Drescher wrote: "RADIA-TOR DAMAGED BEYOND REPAIR AND WAS REPLACED MUST HAVE CHECK VERY SOON TO COVER THIS EXPENSE AND OTHER BILLS SPECIMEN COMING ALONG VERY WELL."

In a letter written two days later, Drescher told Stock: "Of interest is the discovery of ossified [converted to bone] tendons along the neural spines of the dorsal and lumbar vertebrae." However, he took the opportunity to complain: "Perhaps you did not know that we were completely short of funds before we came here this last time. It was necessary for me to borrow twenty-five dollars from Leard to make the trip here, and he is in need of this money at

FIGURE 5.97
As temperatures rise in May, the eight-person, one-dog crew takes a welcome break under the relative cool of the tarp. Photo from the collection of Robert Wallace.

FIGURE 5.98
The forelimbs of the *Saurolophus* are exposed in the center of the upper picture, taken on May 3, 1940. Note the burlap being readied for jacketing between the crew. Photo by Robert Leard; courtesy of the Vertebrate Paleontology Department, Natural History Museum of Los Angeles County.

this time. . . . In addition to this loan from Bob, I was forced to borrow twenty dollars from Miss Smith, and to run a bill of twenty-five or more dollars at the grocery store here."

At about this time Charles McDougall, Tom Fox, and Bob ("Potsey") Wilmoth joined the crew. McDougall (pers. comm. 1998) mentions that there were lots of tarantulas and a few snakes, but it was the scorpions that were a problem. The workers had to empty their shoes each morning. One person didn't and got stung. It was hot, and the digging was dirty work, but still it was fun. As fossils go, the dinosaur was relatively easy to excavate—the bones were much harder than the substrate they were in. At night, the talk around the campfire

FIGURE 5.99
On May 23, 1940, the (upside down) skull of *Saurolophus* has been completely exposed prior to jacketing. Two small faults have offset portions of the specimen. Photo by Robert Leard; courtesy of the Vertebrate Paleontology Department, Natural History Museum of Los Angeles County.

FIGURE 5.100
The skull of the *Saurolophus* is exposed on the far right in this photo taken in July 1940. Meanwhile, the crew begins to jacket the rest of the nearly complete skeleton. Photo by Robert Leard; courtesy of the Vertebrate Paleontology Department, Natural History Museum of Los Angeles County.

was pleasant, and the enormous old army tent was quite comfortable, as the flaps allowed the breeze to come through. Once a week it was necessary to go to Mendota for water, soft drinks, and beer. They traveled in an International Travelall (Drescher had picked it up at the factory), with windows that were like a giant Landrover's. McDougall comments: "The road? from our camp to a real highway was so bad that one of us undergrads rode on the

FIGURE 5.101
Bringing jackets containing
the dinosaur skeleton up the
hill to camp. Photo by Betty
Smith; courtesy of the
Vertebrate Paleontology
Department, Natural
History Museum of Los
Angeles County.

front fender to signal the driver which way to turn to miss especially bad spots. That was a prized assignment!"

In a letter dated May 27, 1940, Drescher pleaded with Stock to come to Panoche to see the exposed *Saurolophus:* "If you could again come to Mendota, I think that you would be repaid for the effort, since now the specimen is completely exposed—a very beautiful sight to a paleontologist. My guess that the specimen would be complete has been proven to be correct. . . . Having entirely delineated the skull and jaws, I will say that it seems very near to *Saurolophus,* there being no hood [crest]. The total length from nose to the end of the tail is twenty-six feet." He also said they could complete work about June 10. Although the hadrosaur did prove to be nearly complete, it was somewhat flattened and a few of the skeletal parts were missing.

While they worked on the *Saurolophus* they were also working on a plesiosaur. On June 12, Drescher told Stock: "We plan to return to Pasadena on Sunday, and will have collected the rest of the Plesiosaur tail from the Panoche Hills." In the same letter Drescher mentioned that Leard had brought a new crew to the Panoche Hills. In July Drescher and Leard were joined by Smith and a new man, John Cushing.

Meanwhile, relations with the Berkeley contingent of paleontologists appear to have been patched up, with some scientific cooperation being undertaken. In a letter of October 1, 1940,

EUSTACE L. FURLONG (1874–1950)

Eustace Furlong was an important figure in the discovery and preparation of fossil Mesozoic reptiles in California for more than fifty years. He grew up in San Francisco, the son of an old California family. He entered the University of California at Berkeley in 1900 and by 1901 had become the vertebrate fossil preparator for paleontologist John C. Merriam. He joined the early fossil reptile collecting trips to Shasta County in 1902 and 1903, so productive for the history of California Mesozoic reptilian paleontology. Here he collected with Annie Alexander, the sponsor of the trips. Although Furlong has only one find solely to his credit, together he and Annie Alexander found forty-three specimens, many of which he actually discovered.

Furlong is largely responsible for the arduous preparation of the reptilian fossils found in Shasta County as well as many later found in Nevada. Chester Stock says of him, "It was Furlong who aided in the elucidation of the fossil record of Upper Triassic ichthyosaurs of western North America by the preparations that were made of the materials for Dr. Merriam." Furlong continued as the vertebrate fossil preparator until May 1910. Three years later he returned to continue his association with paleontology at Berkeley. In 1919

FIGURE 5.102
Robert Leard (left) and Eustace Furlong in the lab at Caltech, ca. 1940.
Photos from the collection of Arthur Drescher.

Charles Camp thanked Stock for "[the] Mosasaur material." And Samuel Welles, too, wrote, on October 11, 1940, asking for information on Stock's plesiosaur material.

A December 11 letter to Stock from Leard—now back with Drescher and John (Cushing?) in the Tumey Hills in the western San Joaquin Valley—mentions a mosasaur find. "Today we prospected and were fortunate enough to locate a string of vertebrae. Thus far we have exposed about 10, some ribs, and two scapula(?). We seem to be going toward the head from just behind the shoulder. I think the find is probably a mosasaur with a fair chance of finding a skull." In a second letter, written December 18, Leard confirmed that there was a well-preserved mosasaur skull attached to the specimen. He then summed up:

> The weather has been bad, cold all last week and rain so far this week, but outlook is good tonight. This has delayed the work to a considerable extent as the road, as you probably realize, is impassable if it is the least bit wet, and the specimen is about two miles from camp. At any rate I expect to have it out by Christmas. . . . On Monday Mr. Drescher secured a position with the U.S.G.S. in San Luis Obispo, so he is no longer with the party. We picked Wahrhaftig up in Fresno on Saturday nite, so our party is now three.

Chester Stock became an instructor of paleontology at UC Berkeley, beginning a long association with Furlong. In 1926 Furlong joined Stock at Caltech. In a letter to Stock dated April 18, 1927, Merriam says, "I have great confidence in Furlong and in the field operations, as you know, he is tactful, uses good judgment, [and is] resourceful and experienced."

Furlong did most of his work in mammalian paleontology, but he again became an important player in Mesozoic reptilian paleontology when Stock sent crews to the Panoche Hills in the late 1930s and early 1940s. Although Furlong did occasionally go into the field, it was his work in the preparation of the fossil marine reptiles and dinosaurs that he should be noted for. Samuel Welles named a plesiosaur *Aphrosaurus furlongi* in honor of Furlong's work at both the University of California Museum of Paleontology and the California Institute of Technology.

Furlong has been described as a quiet man in his later years. In 1945 he retired to Eugene, Oregon, where he continued to work on vertebrate fossils. Late in that year he was struck by a car, receiving injuries from which he never quite recovered. In a letter to Furlong's widow dated February 6, 1950, Stock wrote, "My association with Eustace extended over some of the best years of my life. There was a time when he was almost a father to me, and I shall always hold him warm in my heart for his fundamental goodness and for the much good advice he gave me in the days I knew him the best. . . . He was always a great mentor for the students."

Discoveries and Other Activities: 1940 to 1954

At about this time Welles returned with Max Payne, a geologist, to the western San Joaquin to tie down the locations of previous finds: two mosasaurs *(Kolposaurus* [now *Plotosaurus] bennisoni* and *Kolposaurus* [now *Plotosaurus] tuckeri)* and a plesiosaur *(Hydrotherosaurus).*

Allan Bennison was back in the hills again on August 23, 1940 (Camp 1942), and found an unidentified reptile tooth (UCMP 36252). This unidentified tooth could not be located in the Berkeley collection. In that same year John Hammond (the rancher who had used a hadrosaur bone for a doorstop) found a second ichthyosaur rostrum (beak) in a radiolarian chert cobble, this time near the mouth of Del Puerto Canyon.

A turtle was recorded in the LACM collection in 1941 as having been found in the Panoche Hills by B. Smith (presumably Betty Jean). This date is somewhat problematic, however, as she was applying for residence at Hershey Hall at the University of Southern California in November 1940 and there is no record of her being in the field in 1941. It is probably an earlier find of hers.

On January 7, 1943, Stock wrote to Camp about sending William Otto, Stock's preparator, to Berkeley to discuss plesiosaurs. They were making a full mount of the plesiosaur *Morenosaurus stocki,* and Otto was finding it necessary to restore missing parts, including the skull, so needed to look at the Berkeley material. Otto enjoyed sculpting animals, including extinct fossil forms. He sculpted a nearly life-size *Smilodon* (saber-toothed cat) as well as several scale models of dinosaurs, of which multiples were cast in plastic. He did a wonderful job sculpting the skull of *Morenosaurus stocki* and the missing parts of the mosasaur *Plotosaurus tuckeri,* both of which are on display at the Natural History Museum of Los Angeles County.

The Japanese bombing of Pearl Harbor on December 7, 1941, which brought the United States into World War II, had an impact on paleontology in California. Many of the crew who worked for Stock eventually served in the military, and military security restrictions even got in the way on occasion. On May 4, 1943, Stock wrote to friends about putting together the plesiosaur *Morenosaurus stocki* on the roof of the Mudd Geology Building at Caltech:

> Our plesiosaur mount is reaching an interesting stage. It is so big with its length of 30 or more feet that I am proposing to Furlong and Otto that they make a temporary mount of the specimen on the roof of our building. This will give me an opportunity to take some photographs of the specimen when completed. However the Army men are very anxious to have all loose objects removed from the roof in order to keep the decks clear for possible air raids. If I had the specimen on the roof with its long neck and head projecting over the parapet, the natives down on the street three floor[s] below would wonder what in hell was going on. Perhaps I told you that the Navy moves in on the first of July, and we will have here about 550 men. The Institute hopes to run a regular freshman class along side of the Navy program.

A letter of July 20 mentions that Stock had finished the plesiosaur mount and was starting on the mosasaur.

On August 1, 1943, Stock wrote to a Guy Smoot of Smoot's General Merchandise of Mendota. Stock was contemplating sending Leard—who "plans to have a vacation from his war work"—into the field. He wanted to know if the Panoche Hills had been taken over by the armed forces as a bombing range and said if not he would see if it was "possible to have the institute obtain a special ration book [for food and gasoline] to cover the time of his absence." Smoot replied on August 10: "I am a member of the rationing board at Firebaugh and I took the matter up with the other members—all concur in approving anything you may require for the conduct of your party." Five days later Stock wrote back: "we shall have the necessary gas coupons so that there will be no difficulty from the standpoint of transportation."

On September 26, 1943, Stock wrote Welles congratulating him on his recent publication on plesiosaurs and mentioned that on that day they had just finished the full mount of the plesiosaur at Caltech. He was quite proud of its name, *Morenosaurus stocki.*

In September 1945 Stock made one more attempt to find Mesozoic reptiles in the western San Joaquin Valley, sending John F. Lance to the Panoche Hills. Lance wrote Stock that

FIGURE 5.103
From left to right: William Otto, Eustace Furlong, and Chester Stock make final changes to the plesiosaur *Morenosaurus stocki* on the roof of Caltech's Mudd Geology Building in the summer of 1943. Courtesy of the Vertebrate Paleontology Department, Natural History Museum of Los Angeles County.

FIGURE 5.104
Detail of head of *Morenosaurus stocki* now on display at the Los Angeles County Museum of Natural History. Because the original head was not found, this replacement was sculpted by preparator William Otto. Courtesy of the Vertebrate Paleontology Department, Natural History Museum of Los Angeles County; photo by the author.

he was having his share of car troubles, with a couple of flat tires and ignition problems. Although he was hoping to find a mosasaur, he had had little success finding any fossils. There is no record of any reptilian fossils having been found in the Panoche Hills that year.

Only one fossil find is recorded in the San Joaquin Province between 1943 and 1953. In about 1950 (according to the notes of Welles) John Kepper, a student at Santa Cruz Junior

STOCK'S FIELD CREWS

The crews that worked under Stock in the Panoche Hills comprise just a sampling of the many students he touched. It says a great deal about Chester Stock when we consider the achievements of just a few of them. (See individual biographies for others.)

John Cushing (1916–) received a Ph.D. from Caltech in immunology with a minor in paleontology. He is currently professor emeritus at the University of California, Santa Barbara.

Paul C. Henshaw (1913–1986) did his undergraduate training in economic geology at Harvard and received a Ph.D. in vertebrate paleontology from Caltech. Henshaw taught mineralogy for three years at Caltech and two years at the University of Idaho. He became the CEO of Homestake Mining and later chairman of the board.

Richard Hopper (1914–) received a doctorate in geology from Caltech, where he taught as an assistant in field geology. He later became an oil geologist and in 1939 went to Sumatra, Indonesia, with Chevron. He was working on Timor Island when World War II broke out. Hopper was the first to suggest that there was oil in the Minos area in Sumatra.

Richard (Dick) Jahns (1915–1983) graduated from Caltech with a major in geology and minor in vertebrate paleontology. He worked for the USGS during World War II and later became a full professor in geology at Caltech. He then went on to Penn State University, where he became head of the geology department. From 1965 to 1979 he was dean of the School of Earth Sciences at Stanford University. He was a nationally known authority on earthquakes and plate tectonics.

Clyde Wahrhaftig (1919–1994) received his bachelor's degree in geology from Caltech in 1941 and a Ph.D. in geology from Harvard in 1953. He then worked for the USGS until his death, but for twenty-two of those years he also taught at the department of geology and geophysics at UC Berkeley and was an emeritus professor of geology at Berkeley. An outstanding geologist, he was awarded the Distinguished Career Award by the Geological Society of America.

College (now known as Cabrillo College), found and kept three mosasaur teeth. A plesiosaur vertebra was found by Mrs. J. Ray in 1954.

Welles and His Students: 1955 to 1959

In October 1955 Welles returned to the Panoche Hills with several students in tow (including James Guyton, author years later of *Glaciers of California* [University of California Press, 1995] and today professor emeritus at California State University, Chico). Welles said in his field notes that the area was hard to recognize, but they found three badly weathered mosasaur vertebrae. In 1956 Welles and his students were back in the Panoche Hills, where John Cosgriff, Miss West, and Hugh Hildebrandt each found mosasaur remains, and students identified only as Sheldon and Hosley got credit for two more. That made for a total of eight mosasaur remains discovered.

The following March Welles was back in the Panoche Hills with his paleontology class, hav-

There is a proposal to rename the Pleistocene lake that formed the Corcoran Clay and filled much of the San Joaquin Valley "Lake Clyde" (Sarna-Wojcicki 1995). It would be a fitting tribute to Clyde Wahrhaftig, as "Lake Clyde" once lapped against the eastern edge of the Panoche Hills where he began his career as a geologist excavating Mesozoic reptiles.

Robert E. Wallace (1916–) earned a doctorate in geology at Caltech. He later taught geology at Washington State in Pullman, then joined the USGS at Menlo Park and became an expert on the San Andreas Fault.

Robert W. Wilson (1909–) was a graduate teaching assistant for Chester Stock at Caltech, where he received his doctorate in geology and paleontology. He then held a Sterling Research Fellowship at Yale University. At the University of Colorado at Boulder he was appointed to fill a vacancy in vertebrate paleontology and geology. Today, in his nineties, he works in paleontology at the Museum of Natural History at the University of Kansas.

One person who virtually disappeared after her work in the Panoche Hills was Betty Jean Smith. I have spent many hours trying to find out what happened to her. Smith found one of the best-preserved California dinosaurs—a *Saurolophus*—plus a mosasaur and a turtle. Stock wrote of her on December 5, 1940: "Betty Jean Smith comes from a cultured and fine family in Pasadena. She is a girl of good principles and character and possesses withal a very pleasing personality." Drescher (pers. comm. 1998) said, "She had a deep interest in paleontology and by persistence she prevailed on Dr. Stock to let her join the fossil collecting crew. She was a hard working, energetic prospector and collector. She did her full share of the work and camp duties often under hot insectiferous rough conditions and she had the respect of all of the crew. There was never any awkwardness because of her sex in an otherwise all-male crew." It is my hope that someone will name a new species of California Mesozoic reptile after Smith.

FIGURE 5.105
"Baby" mosasaur
(*Plotosaurus*) on display at
Monterey Peninsula College.
Photo by the author.

FIGURE 5.106
Carl Zarconi preparing
bones of the plesiosaur
Morenosaurus stocki in
the lab of Modesto Junior
College. Photo from the
collection of Mervin
Lovenburg.

ing arranged to meet some students of Professor Haderlie's from Monterey Peninsula College. They were led by "Digger Don" Howard, a rather free spirit with a passion for digging fossils (Mel Bristow, pers. comm. 1999). Besides plesiosaur remains that Howard found in the Panoche Hills, the trip produced incomplete remains of two other plesiosaurs and three mosasaurs.

In May 1958 Welles's paleontology class was again in the Panoche Hills, this time joined by Frank Bush, an instructor at Sacramento Junior College. The group found a "good mosasaur skull & neck." The next March Digger Don hit pay dirt once again when he discovered a more or less complete skeleton of a "baby" mosasaur (it may have been just a small form). Howard prepared the specimen, and Welles agreed that it should go on display at Monterey Peninsula College, where it remains today.

Various Discoveries: 1960s

In 1963 S. D. Webb found a mosasaur tail while in the Panoche Hills with Joseph P. Gregory of the University of California, Berkeley. In March 1964 four people (Peck, Nelson, Allison, and Swan) discovered some sort of reptile tooth in Contra Costa County. In May of that same year Mervin Lovenburg went into the hills southwest of Modesto with his paleontology students from Modesto Junior College. One of his students, Carl Zarconi, thought he had found something that looked like a shark tooth projecting from a lump of rock. He brought the lump to Lovenburg, who recognized that the lump was actually a single large reptile vertebra; the "shark tooth" turned out to be a tooth from a plesiosaur. Lovenburg contacted Welles, and on June 24, 1964, a team led by Welles and Gregory arrived to excavate more of the creature. Welles, Gregory, Ed Mitchell, and a team of students finished the excavation October 23–26. They recovered nearly the entire rear half of the beast. Lovenburg jokingly told Carl that he wanted to name the specimen *Modestoenses zarconei,* but as it turned out, it was another *Morenosaurus stocki.*

The Staeblers: 1975 to the Present

For eleven years no new finds were reported from the western San Joaquin Valley. Then in March 1975 a new phase of paleontologic discovery began when Chad A. Staebler found mosasaur remains in the Panoche Hills. Chad went there with his father, Arthur Staebler, a professor at Fresno State College (now California State University, Fresno) and eight of his students to prepare for a trip to the Chinle Formation in Arizona. Chad knew that fossil reptilian remains had been found in these hills because a cast of *Hydrotherosaurus alexandrae* was mounted in McLane Hall at Fresno State. After Chad found mosasaur vertebrae he urged his father to take his natural history classes to the Panoche Hills to trap small mammals; that way, Chad had legitimate access to the area (since a permit was required), and while the students did biology he could look for fossils. A bright young student by the name of Dianne Yang came along on some of these early trips. Not only did she make many important discoveries, but in 1983 Dianne capped off her interest in paleontology with a master's thesis, "A Study of the Pectoral and Pelvic Appendages of California Mosasaurs." She joined the Staebler family when she and Chad married.

Now concentrating on paleontology rather than biology, Arthur Staebler's crews went back to the Panoche Hills in January 1976 to look at a site where Welles had found one of the big plesiosaurs. There Chad discovered postcranial material from another mosasaur. Three years later, in the same hills, he discovered part of another mosasaur as well as the remains of two rare turtles, *Osteopygis* and *Basilemys.*

Nineteen-eighty was a banner year for Chad: he discovered two more turtles and the incomplete remains of twenty different mosasaurs. In one weekend alone the Staebler crew found ten mosasaurs, one of which was the second known specimen of *Plotosaurus bennisoni.* It was complete with skull, entire vertebral column, the pectoral girdle (shoulder bones), and some elements of a pectoral appendage (arm). Welles came out to help put a plaster jacket on the skull, and the following weekend Chad and Welles brought it to Fresno State. The site was on the crest of a hill; it was a day's work lowering the fossil in its plaster jacket down into the gully, then lugging it up the hill on the other side. Another mosasaur they found on that same day was oriented straight into the ground, and they could only retrieve the

ARTHUR E. STAEBLER (1915-)

The knowledge, combined with the organizational skills, of biologist Arthur E. Staebler has resulted in more discoveries of Cretaceous reptilian remains in California than any other person. Staebler received a bachelor's degree in 1938, an M.S. in zoology in 1940, and in 1949 a Ph.D. in ornithology, all from the University of Michigan. He was director of the W. K. Kellogg Bird Sanctuary of Michigan State University from 1948 until coming to Fresno State College in 1955 to teach biology.

In 1975 Arthur Staebler began taking his vertebrate natural history and ornithology classes to the Panoche Hills on the west side of the San Joaquin Valley, in part on the urging of his son Chad, who knew of reptilian fossil discoveries in the area. In that same year Arthur began to teach vertebrate paleontology and did extensive field collecting of fossils in the Panoche Hills for Fresno State. Although Chad found most of the fossils, it was Art Staebler who was the heart and soul of the expeditions, doing the organizing, instruction, and much of the science. Most of all, he was responsible for the high morale of his students in the field, without which the harsh conditions in the Panoche Hills would have quickly spelled doom for any successful fossil hunting.

FIGURE 5.107
Art Staebler in 1998 with plesiosaur paddle from the Panoche Hills.
Photo by the author.

FIGURE 5.108 (LEFT)
Chad Staebler (with shovel), Nancy Muleady-Mecham, and other Fresno students at the excavation site of a mosasaur, around 1975. Photo courtesy of the California State University, Fresno, Department of Biology.

FIGURE 5.109 (RIGHT)
Art Staebler with carapace and bones of the terrestrial(?) turtle *Basilemys.* Photo courtesy of the California State University, Fresno, Department of Biology.

Phil Desatoff, one of Staebler's former students and now a geologist with Fresno County, said, "Finding and digging fossils in the heat of the Panoche Hills is not what most people would call fun. Art's captivating stories and group camaraderie made the trips fun and interesting on their own. His exuberance and knowledge were amazing. His students could not keep up with him, and even today, in his mid-eighties, he's got more energy than most of them."

In 1989 Art arranged for an exhibit of the Panoche fossils at the Fresno Metropolitan Museum, inviting paleontologist Jack Horner as keynote speaker. He also brought Walter Alvarez to the Panoche Hills to search for the K-T boundary—and they found it, too. On an ongoing basis, Art Staebler worked closely with Berkeley scientists like Samuel P. Welles to make sure that Chad's finds would be curated properly and his discoveries published for science.

Together, Art and Chad Staebler surveyed forty-four thousand acres of hilly ground on the west side of the San Joaquin Valley. Then in 1981, after massive research, Arthur wrote *Survey of the Fossil Resources in the Panoche Hills and Ciervo Hills of Western Fresno County, California.* The voluminous work outlined in detail for the first time the vast fossil resources of the area.

Art Staebler still occasionally works in the heat of the Panoche Hills and continues to be an inspiration to fossil hunters.

posterior portion. The Berkeley crew came out and excavated another four or five feet of vertebral column but called it quits. The rest is still there.

Art Staebler's students made several discoveries as well. Stephen Lozano found a skull-less mosasaur specimen that curved and doubled back on itself. Mark Yarborough encountered more mosasaur remains, and Paul Dake found the first partial pelvis of a mosasaur in California (C. Staebler 1981). Woodrow ("Woody") Moise discovered portions of a turtle plastron (the underside of the shell) and paddle as well as fifteen articulated vertebrae of a mosasaur tail section. He later helped the Staeblers map out the Moreno Formation and pin down the age and stratigraphic relationships of the fossils.

In that same productive year, 1980, Mark Goodwin of the UCMP recovered a four-foot-long vertebral column of a mosasaur with a shark tooth on the underside. In 1981 another Staebler student, Philip Desatoff, found two mosasaur paddles, while Chad discovered the incomplete skeletons of six other mosasaurs. On one trip Marsha Downing, Desatoff's future wife, was sitting with the group lamenting that she hadn't found a thing. But when she got up Marsha found that she had been sitting on a mosasaur vertebra!

On June 13, 1981, with TV crews on hand, Welles and Lewis Bremer IV met with the Staebler group in the Panoche Hills. Yang had found a nearly complete front portion of a mosasaur, including part of the skull.

CHAD STAEBLER (1957–) AND DIANNE YANG-STAEBLER (1954–)

Chad Staebler started collecting minerals and rocks even before he had entered grammar school. While studying geology at California State University, Fresno, he noticed the large cast of the plesiosaur *Hydrotherosaurus alexandrae* from western Fresno County on display. This prompted him to convince his father, a biology professor at the university, to lead field biology classes in the Panoche Hills so he could go along and search for fossils. In 1975, after Chad Staebler made his first important fossil finds, the trips became dedicated vertebrate paleontology excursions. Chad met Dianne Yang, his future wife and collaborator, on one of these trips. After finding mosasaurs of her own, Dianne Yang-Staebler became so interested in fossils that she earned a master's degree at Fresno in 1983 with her thesis "Study of the Pectoral and Pelvic Appendages of the California Mosasaurs."

Chad Staebler was the spark that for many years kept the fossil flame lit

FIGURE 5.110
Chad Staebler in 1980 proudly displays a mosasaur (encased in burlap and plaster) that he has just recovered from the Panoche Hills. Photo courtesy of the California State University, Fresno, Department of Biology.

DINOSAUR POINT

Dinosaur Point on San Luis Reservoir near Los Banos in Merced County has been an interesting source of stories about dinosaurs and other Mesozoic reptiles. I once heard of a hadrosaur having been found at Dinosaur Point, and I was even told that a dinosaur was discovered here and was now housed at Stanford University. On a field trip with a prominent geological association, I noticed an entry in the field

FIGURE 5.111
Sign for Dinosaur Point on Highway 152 at San Luis Reservoir. Photo by the author.

in the hills west of Fresno. He is one of those extremely rare people with an almost supernatural intuition for locating fossils. As his friend Woody Moise put it, "It was as if Chad could smell fossils." After joining the Staeblers on their digs, Samuel Welles recognized this talent and encouraged Chad to participate in vertebrate paleontologist Reid MacDonald's paleontology trips. Chad spent three successive summers on MacDonald's digs, finding ten or twelve fossils to the other students' two or three, including a hadrosaur vertebra in Montana, part of a *Stegosaurus* fin in Utah, and in the Bighorn Mountains of Montana he found sauropod limb elements. MacDonald, too, became convinced that Chad had fossil radar. On a university field trip to South Dakota, MacDonald was looking for fossils of *Hesperornis* (a toothed diving bird). Having no success, he proposed that the first person to find a *Hesperornis* would get an A for the course. Chad promptly walked no more than five feet from MacDonald and pointed to one. Chad got his A, but he more than earned it.

Over the years Chad's California finds have included the remains of two turtles, a plesiosaur, a dinosaur, and forty-six mosasaurs. Chad Staebler has found more fossil Cretaceous reptiles than any single person in California.

guide prepared for the trip about a mosasaur having been found at Dinosaur Point. I asked the author and trip leader what his source for this information was, and he replied that a student told him that the "Fresno people" had found it. I asked the "Fresno people"—Art, Chad, and Dianne Staebler—and, as I suspected, they knew nothing of the creature. They suggested that the point got its name from its resemblance to a dinosaur. I called the folks at the visitors center and the park rangers at San Luis Reservoir, and they said the same. All they could say was that the outline of the point, or the shape of the hill, or perhaps the shape of a nearby lake looked like a dinosaur.

Old tales rarely die, but there are no dinosaur remains from "Dinosaur Point."

FIGURE 5.112 (ABOVE)

From left to right: Art Staebler, Sam Welles, Chad Staebler, and Dianne Yang-Staebler inspect mosasaur flipper bones as a TV crew films, June 13, 1981. Photo courtesy of the *Fresno Bee,* Fresno, California.

FIGURE 5.113 (TOP RIGHT)

Dianne Yang-Staebler (straw hat, foreground), Chad Staebler (leather hat), Sam Welles (with binoculars), Art Staebler (with brush), Steve Ervin (kneeling), and students inspect a mosasaur, June 13, 1981. Photo courtesy of the *Fresno Bee,* Fresno, California.

FIGURE 5.114

Lower jaw of hadrosaur found by Chad Staebler in 1982. Courtesy of the University of California Museum of Paleontology; photo by the author.

It wasn't until 1982 that Chad recovered his first dinosaur remains, consisting of vertebrae and ribs as well as the beautiful lower jaw of a hadrosaur. That same year he encountered the incomplete remains of four more mosasaurs, and student Paul Dake discovered a mosasaur pelvic paddle.

In the late afternoon of May 2, 1983, Chad was standing on a hilltop in the Panoche Hills looking for fossils when he heard a rumbling sound. He looked out to see the earth moving

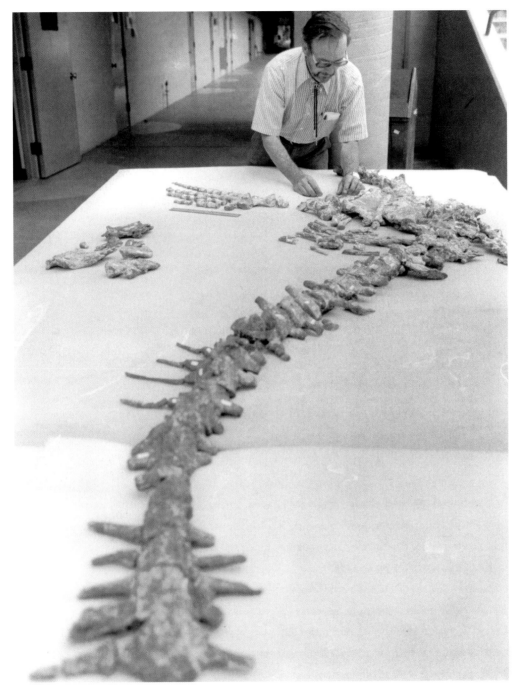

FIGURE 5.115
Art Staebler with mosasaur skeleton laid out in a Fresno State
lab. Photo courtesy of California State University, Fresno,
Department of Biology.

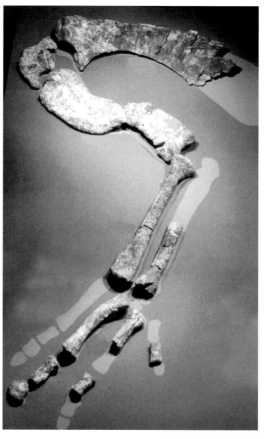

FIGURE 5.116 (ABOVE)
Coalinga residence after the magnitude 6.5 earthquake of May 2, 1983. Photo by the author.

FIGURE 5.117 (RIGHT)
Hadrosaur limb bones discovered by Steve Ervin in 1985. Photo courtesy of California State University, Fresno, Department of Biology.

in giant waves, like some gargantuan ship had just passed and the wake was coming toward him. He saw the waves approach and then felt the ground undulate as they passed beneath him. It was the 6.5 Coalinga quake, and the epicenter was just a few miles to the south. Much of Coalinga was destroyed in this event.

In 1984 Chad found the partial remains of two mosasaurs, and Art Staebler discovered the shoulder girdle and upper appendages of another *Plotosaurus*. In 1985 Chad happened upon the partial remains of two more mosasaurs, and Art a paddle with a claw. The most exciting find that year came when a fellow professor at Fresno, Steven Ervin, discovered more hadrosaur remains: forelimbs, metacarpals (upper fingers), two femora (upper legs), vertebrae, a tibia (lower leg), and a pectoral (shoulder) girdle.

In 1987 Chad discovered remains of three different mosasaurs. The next year Fresno State had an exhibit of some of the Staebler crew's fossil finds from the western San Joaquin. Paleontologist John (Jack) Horner was invited to attend, and he also joined the Staeblers on a

FIGURE 5.118 (LEFT)
Paleontologist Jack Horner found plesiosaur remains in the Panoche Hills. Photo courtesy of Jack Horner and the Museum of the Rockies.

FIGURE 5.119 (ABOVE)
Stomach stones (gastroliths) from a Panoche Hills plesiosaur. Courtesy of the Vertebrate Paleontology Department, Natural History Museum of Los Angeles County; photo by the author.

fossil hunt in the Panoche Hills. There Horner discovered the vertebra of a plesiosaur, the only fossil recorded from the western San Joaquin that year.

In 1989 Chad went out to try to find something to videotape for an exhibit of Panoche Hills fossils at the Fresno Metropolitan Museum when he happened upon the skeleton of a plesiosaur (elasmosaur) complete with gastroliths (stomach stones). The gastroliths were unlike the rocks of the Moreno Formation, in which the fossil was found, but more like those of the Franciscan Formation cherts in the Coast Range or the foothills of the Sierra. This indicates that these animals may have traveled some distance to select proper stones, a phenomenon that Williston (1902) documented for interior seaway plesiosaurs. (Most likely, these gray cherts came from Sierran sources, since the Coast Range probably didn't exist at that time.)

Chad did not return to the field until 1991, when he found two more partial mosasaurs and a small plesiosaur. The following year J. Howard Hutchison, a fossil turtle expert and former museum scientist of the vertebrate collection at the UCMP, discovered a mosasaur vertebra.

Over the years Chad noticed that several of the mosasaurs he had found were aligned in a more or less northeasterly direction, leading him to speculate that there may have been an ancient ocean current in that direction.

The rate of fossil reptile discovery slowed after 1991, but the Staeblers still occasionally

FIGURE 5.120
From left to right: Art
Staebler, Chad Staebler, and
Dianne Yang-Staebler exca-
vate a plesiosaur in the
Panoche Hills in the 1990s.
Photo courtesy of the
California State University,
Fresno, Department of
Biology.

ventured into the Panoche Hills. In 1995 they joined geologists Walter Alvarez of the University of California and Jan Smit of the Free University of Amsterdam to search for the K-T boundary layer in the Panoche Hills. Alvarez (pers. comm. 2002) credits Smit with having done more to advance the K-T impact theory—originally proposed by Walter and his father, Luis Alvarez—than any other scientist. According to this largely accepted theory, a meteor impact contributed to the extinction of many life forms on Earth 65 million years ago, at the K-T boundary, that is, the boundary between the Cretaceous (K) and Tertiary (T) Periods. All of the dinosaurs, all the large marine reptiles (excluding turtles), and all the pterosaurs went extinct at this juncture, along with many other plants and animals.

The Staeblers had previously taken Alvarez to the Panoche Hills in March 1984. On the 1995 visit, however, they noticed that a portion of the K-T boundary manifested itself rather unusually in the region: the associated strata contained weathered glass spherules, some as big as buckshot (W. Alvarez, pers. comm. 2002). Although glass spherules have been found elsewhere in K-T boundary sediments, they are usually of microscopic size. They may represent rocks that were liquified into droplets when the meteorite struck, and then blasted skyward at the Yucatán (Mexico) impact site. They then cooled and solidified as they settled

FIGURE 5.121
Dirt flies at a Staebler dig.
Photo courtesy of California
State University, Fresno,
Department of Biology.

back to Earth. Perhaps the reason the glass spheres are larger at the Panoche Hills is that the locale is closer to the Yucatán than are most other sites revealing the K-T boundary.

In 1997 Art, along with Carlos Fonseca of Stanford University, went into the Panoche Hills to work on plesiosaur bones Fonseca had discovered. Art was in the field again in June and October 1998 continuing the excavation. He said he found more gizzard stones and that he still has three more reptile specimens (two mosasaurs and a plesiosaur) to excavate.

According to Art Staebler there is much work to be done in the western San Joaquin. He pointed out to me the areas that have been searched and the likely areas that have not. The unsearched areas are more extensive than those searched. Additionally, erosion is constantly exposing new fossils. Many new and exciting finds are yet to come from these hills. Perhaps one day Arthur Staebler's grandchildren (Chad and Dianne's kids) will join in on the Fresno fossil-hunting tradition that began with the 1937 discovery of *Hydrotherosaurus alexandrae.*

COAST RANGES PROVINCE

The Coast Ranges Province of California runs roughly from just north of Santa Barbara to the Oregon border. Prior to our understanding of plate tectonics the source and structure of the older rocks of the Coast Ranges were almost a complete mystery. Today we know that most of these rocks are either from Sierran bedrock rafted north by the San Andreas Fault, or rocks that originated in a trench or an open-ocean deepwater setting. The deepwater rocks

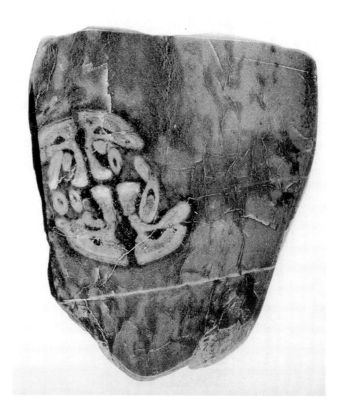

FIGURE 5.122
Cross-section of ichthyosaur
rostrum found in radiolarian
chert. From Camp 1942b.

that today make up the Franciscan Formation were later squeezed and accreted to this part of the world through the convergence of the North American Plate and the oceanic Farallon Plate to its west. These rocks have undergone significant deformation. Outliers of Great Valley Group rocks deposited nearer to the shore are also present.

Two of the three specimens of Mesozoic reptile remains from the Coast Range rocks were discovered in radiolarian cherts: highly contorted, thinly layered beds originally deposited in the deep ocean well offshore from the continent. They are made primarily of the silica-rich skeletons (tests) of microorganisms called radiolaria. Most of the dating of the Mesozoic sedimentary rocks in the Franciscan Formation has been done using radiolaria fossils (see figs. 1.16 and 1.17).

Of the three sets of Mesozoic reptilian remains found in the Coast Range rocks of California the first is a Jurassic ichthyosaur rostrum (beak). It was discovered in the summer of 1935 in the Corral Hollow drainage of San Joaquin County by Neal Johnstone Smith while he and Allan Bennison were on a lunch break from geological fieldwork (Bennison, pers. comm. 2000). The fossil was in a piece of radiolarian chert that had been redeposited in

FIGURE 5.123
Olsonowski's plesiosaur vertebra centrum. Courtesy of the University of California Museum of Paleontology; photo by the author.

Pleistocene (Ice Age) gravels and subsequently washed into the Great Valley Province from the Coast Ranges.

The second find, in 1940 in Stanislaus County, was the front part of a skull of a Jurassic ichthyosaur, discovered by rancher John Hammond, who liked to collect odd rocks (and who, as we have seen, used a hadrosaur leg bone as a doorstop). This fossil, too, was in a piece of chert that, though found in redeposited material in the Great Valley Province, had its origins in the Coast Ranges.

In 1949 a third discovery was made, this time in San Luis Obispo County, when Betty Olsonowski, on a field trip led by paleontologist J. Wyatt Durham of the University of California (who found the first dinosaur remains in Baja), uncovered two vertebrae of a plesiosaur (see fig. 5.123). Other than a specimen found by Thomas Knock in the Great Valley Group of the western Sacramento Valley (not previously recognized as a plesiosaur), this was the first evidence of a Jurassic plesiosaur discovered west of the Rockies. Olsonowski's fossil is the only Mesozoic reptile to have been found right in the Coast Ranges Province. It came from a limestone concretion weathered out of what Durham called Franciscan-Knoxville shales.

All three of these creatures must have died with their carcasses decomposing in the open ocean. Their bones then rained down onto the ocean floor, where they became trapped in sediments. We can only speculate whether these animals frequented the open ocean or their carcasses simply floated out from a nearshore environment.

Although not many Mesozoic reptile fossils have been found in the Coast Ranges Province, there is further potential. A search of the Franciscan Formation cherts upstream from the two ichthyosaur finds seems in order.

6

THE SOUTHERN PROVINCES

THE SOUTHERN PROVINCES begin just north of the Los Angeles Basin in the Transverse Ranges Province, an area with relatively few important Mesozoic reptilian discoveries. To the east of the Transverse Ranges Province lies the Mojave Desert Province, where dinosaur trackways in the Aztec Sandstone provide a glimpse into the lives of desert-dwelling dinosaurs. The bulk of this chapter focuses on one of the most productive and exciting areas of discovery, the Peninsular Ranges, a series of mountains that run from the Los Angeles Basin to the tip of Baja California. Here we find tales of discovery and paleontological struggle that rival the early days of dinosaur exploration in the middle of the continent.

TRANSVERSE RANGES PROVINCE

The Transverse Ranges Province consists of mountains that run at an angle (transverse) to the rest of the mountains of California, that is, in an east-west direction. Much of this province lies to the north and east of the Los Angeles basin. Geologically, it is extremely complex.

The Cajon Pass area, where highway I-15 passes into the Mojave Desert, has been the site of discovery of the partial remains of two elasmosaurid plesiosaurs. The first single vertebra described was found in 1978 by Royce Colman of UC Riverside in marine rocks of the San Francisquito Formation just east of the San Andreas Fault. The rocks had been mapped as Paleocene in age (ca. 60 MYBP), although it is doubtful that plesiosaurs survived past the K-T boundary (65 MYBP). This vertebra may have eroded out of older rocks, then been deposited in the younger sediments (if in fact these rocks are Paleocene).

In 1976 Tom Greer, a volunteer from the San Bernardino County Museum, had also

FIGURE 6.1
Tracks of dinosaurs in the dune sand of the Jurassic Mojave Desert.

made a discovery in the same area, again in the San Francisquito Formation: portions of about forty plesiosaur vertebrae, of which ten were relatively complete (Lucas and Reynolds 1993). As these ten vertebrae were all in their original life position, it is unlikely that they were reworked from older beds. This means that either part or all of the Cajon portion of the San Francisquito Formation is Late Cretaceous in age, or at least one species of plesiosaur survived the K-T boundary (Robert Reynolds, San Bernardino County Museum, pers. comm. 1999).

MOJAVE DESERT PROVINCE

The Mojave Desert Province lies inland of the mountains that run the length of southern California. The mountains cause the moist Pacific air to rise, and as the air rises it expands and cools, often leading to the production of clouds and precipitation. On the leeward side of the mountains the now relatively dry air falls and compresses, thus heating up. This leaves the area behind the mountains of California in what is called a rain-shadow desert. Compared to the desert of the Basin and Range Province, the Mojave is hotter in the summer and drier because it is farther south and generally at lower average elevations.

Although no Mesozoic reptilian remains have yet been found in this province, other important evidence survives in the form of several dinosaur footprints and trackways discovered by James R. Evans (1958) in the Early Jurassic Aztec Sandstone of San Bernardino County. They are the only known dinosaur tracks found in California, and to date the earliest evidence of dinosaurs in the state.

Robert E. Reynolds (1983, 1989) of the San Bernardino County Museum analyzed the tracks and says they come from three different small carnivorous dinosaurs. One is similar to *Grallator,* another to *Anchisauripus,* and the third remains unidentified. The tracks are impressed in lithified, cross-bedded dune sand, indicating that these dinosaurs probably roamed an interior coastal desert. These dunes were most likely formed from sand blown inland from the shore of an interior seaway that, in the Cretaceous, divided the eastern from the western portions of North America. Reynolds (1989) reports that the steepness of the track impressions suggests that many were made in damp sand. The Namib Desert of Africa, the Atacama Desert of South America, and even the coastal deserts of Baja California are commonly visited by fogs that dampen the sand dunes. This may also have been the case for the Jurassic Mojave Desert Province.

Searching and perhaps splitting the Aztec Sandstone has the potential to yield more evidence of Jurassic reptiles in California.

FIGURE 6.2
Bob Reynolds with Early
Jurassic dinosaur tracks on
display in the San
Bernardino County
Museum. Photo by the
author.

PENINSULAR RANGES PROVINCE: ORANGE AND SAN DIEGO COUNTIES

The Peninsular Ranges Province of California comprises the area west of the Salton Trough and south of the Los Angeles Basin. It is named for the Baja Peninsula, which extends to the tip of Baja California at Cabo San Lucas and is bordered by the Pacific Ocean on the west and the Gulf of California on the east. Although this book is about the Mesozoic reptiles of California, geology and paleontology do not stop at political boundaries. Discoveries that have occurred in the Mexican portion of the Peninsular Ranges Province have a bearing on the fossils in the rest of California (which I refer to in what follows as Alta, or upper, California).

The entire Peninsular Ranges Province has moved north-northwest due to plate tectonic actions of a ridge in the Gulf of California and associated transform faulting. The Baja Peninsula is being rifted away from the mainland of Mexico in much the same way the Red Sea is widening. The northernmost transform fault—the San Andreas—has caused the entire

FIGURE 6.3
Late Cretaceous Peninsular Ranges Province with a braided stream being crossed
by a lambeosaur and an ankylosaur (*Aletopelta*), as a crocodilian looks on.

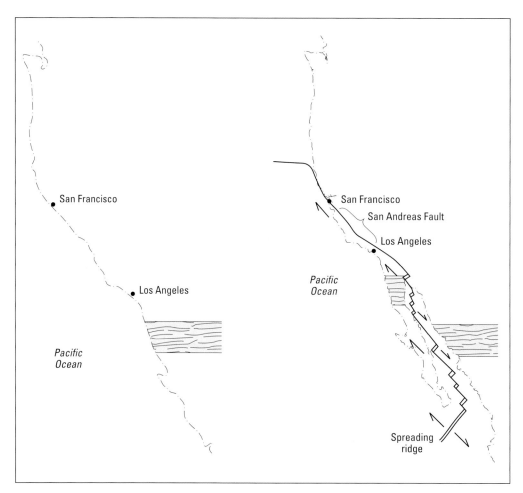

FIGURE 6.4
Map showing the northwestward movement of the Peninsular Ranges
Province relative to most of California. Stippled area indicates the lati-
tude of Mesozoic fossil reptile localities discovered in that province.

peninsula to also move in a northwestward direction about three hundred miles. This means
that all of the Mesozoic reptile fossils of the Peninsular Ranges Province, including the ones
in Alta California, are actually from latitudes farther south. This makes the gap between the
fossils found in more northerly provinces bigger and therefore these paleoenvironments cli-
matologically more disjunct.

In the Peninsular Ranges Province of Alta California reptilian remains have been found in
coastal areas in both Orange and San Diego Counties, in the Mesozoic marine rocks between
the Santa Ana Mountains and the Pacific Ocean.

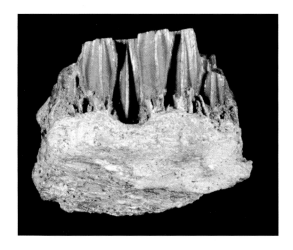

FIGURE 6.5 (ABOVE)
Vertebrae from a Jurassic? plesiosaur found by
G. P. Kanakoff—perhaps the oldest Mesozoic
reptilian remains from the Peninsular Ranges
Province. Courtesy of the Vertebrate Paleontology Department, Natural History Museum
of Los Angeles County; photo by the author.

FIGURE 6.6 (LEFT)
Hadrosaur maxilla (upper jaw) with teeth, discovered by B. Moore in the Cretaceous Ladd Formation of Orange County. Courtesy of the Vertebrate Paleontology Department, Natural History
Museum of Los Angeles County; photo by the
author.

The records of the Natural History Museum of Los Angeles County show that G. P. Kanakoff found a string of elasmosaur (a long-necked plesiosaur) vertebrae in the Jurassic (?) Bedford
Canyon Formation of Orange County. These are the oldest Mesozoic and the only Jurassic
reptilian remains from the Peninsular Ranges. No date for this find is listed in the records.

Another undated find is that of a hadrosaur maxilla (upper jaw) with teeth, discovered
by B. Moore in the Cretaceous Ladd Formation of Orange County. In 1950 a plesiosaur centrum (spool of the vertebra) was found in the same formation by Marlon V. Kirk.

Riney, Drachuk, and Case: 1967 to 1982

Many successful fossil hunters develop their skills at an early age. In 1967 Brad Riney, only
thirteen and still in junior high school, went looking for ammonites in the sea cliffs at La
Jolla. Entering a sea cave, he found a rounded concretion that, having eroded out of the
cliffs, was now a cobble on the beach. The fossil bone sticking out of it turned out to be
his first dinosaur fossil: a single hadrosaur neck vertebra from the Late Cretaceous Point
Loma Formation.

FIGURE 6.7
The sea cliffs of San Diego County have yielded important Mesozoic
reptilian remains. Photo by the author.

In 1978, also while in junior high and out looking for ammonites, Robert Drachuk collected a hadrosaur cranial fragment from a limestone concretion in the Late Cretaceous Ladd Formation of Orange County. Drachuk donated it to the Orange County Natural History Museum.

In about 1980 a thirty-two-year-old surfer and diver, Leon Case, while walking a San Diego County beach, discovered the midsection of a right dentary (lower jaw) of a hadrosaur, complete with teeth, in a beach cobble. It was from the Cretaceous Point Loma Formation. This same formation proved fruitful again when in February 1982 Riney, also on a walk along a rocky beach in San Diego County, found a concretion (now an eroded cobble) that contained two caudal (tail) vertebrae of a mosasaur (a large seagoing lizard). He later found more mosasaur remains in the cliffs to the south. Riney's skill at finding important fossils was recognized by Tom Deméré of the San Diego Natural History Museum, who hired him as a curatorial assistant and field paleontologist for the museum.

Riney and Cerutti: 1983 to 1986

It was at about this time, with the advent of environmental impact reports (EIRs), that paleontologist Mark Roeder helped develop a program to salvage fossils exposed during exca-

FIGURE 6.8 (LEFT)
Robert Chandler holds a
hadrosaur femur discovered
by Bradford Riney in siltstones at Carlsbad,
California. Note fossil oysters still attached. Photo
courtesy of the San Diego
Natural History Museum.

FIGURE 6.9 (RIGHT)
Richard Cerutti at the site
of his 1996 mosasaur find.
Photo by the author.

vation in the numerous construction sites that were blossoming in the southern California area. He and his colleagues set out to make paleontological resources, which are inherently nonrenewable, part of EIRs, with provisions for monitoring and subsequent collection, excavation, and curation of fossils where necessary. These efforts have not only saved a tremendous amount of fossils but also provided extensive information about the paleontological prehistory of the area.

In December 1983 Riney found the femur of a hadrosaur exposed by a bulldozer during grading operations in the Point Loma Formation in the town of Carlsbad. Attached to it were fossil oysters that had grown on it as it lay on the shallow, silty seabottom. Fellow paleontologist Richard Cerutti (also of the San Diego Natural History Museum) helped excavate the specimen, then back in the lab Riney prepared it from the surrounding rock.

In June 1986 Cerutti was walking along the cliffs of Point Loma at low tide when he found mosasaur remains—the midsection of a lumbar vertebra as well as thoracic vertebrae with ribs—in Late Cretaceous siltstones. Demére, Riney, and paleontologist Richard Estes went to the site to assist in the excavation. The bones were taken to the museum where Cerutti

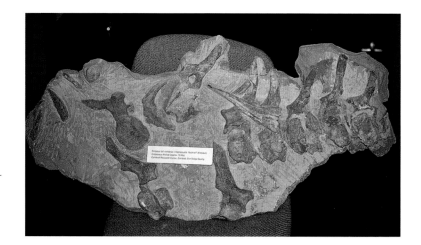

FIGURE 6.10
Thirteen cervical (neck) vertebrae of a hadrosaur found by Bradford Riney in San Diego County. Photo by the author.

prepared them and where today they are on display. When I was in San Diego in 1998 Cerutti and Riney took me to the Point Loma site, where I was shown a few bone fragments still protruding from the sea cliff; however, they are too dangerously situated to be extracted.

In December 1986 Riney found thirteen cervical (neck) vertebrae of a hadrosaur in the Cretaceous Point Loma Formation of San Diego County. He was following a bulldozer in a cut that exposed a concretion with a bone sticking out. Riney prepared the specimen for the museum. Shortly thereafter he made another find: a small foot bone from an unidentified dinosaur.

Riney's Ankylosaur: 1987

In May 1987 Riney made his most exciting dinosaur discovery to date, of a dinosaur that had never been seen in California: the fairly complete postcranial skeleton of an ankylosaur, a medium-sized, plant-eating armored dinosaur. While working on a paleontological monitoring project for a street extension in Carlsbad, Brad noticed a resistant area of mudstone in a trench. Upon inspection he found that it had bones sticking out, and they weren't just isolated bones. Exposed along a twelve-foot length, they appeared to be a skeleton. A tooth proved that it was indeed a dinosaur; more important, it wasn't a hadrosaur, so Riney knew he had a rare find.

The bones were brought to the attention of Deméré, who helped organize the field excavation and later published three papers on the creature. The excavation was difficult and required the help of other paleontologists, including Roeder. After a week of toil in which fifteen feet of overburden were removed, they were able to excavate what was left of the animal. During the original excavation, unfortunately, heavy equipment had lifted material from

FIGURE 6.11 (LEFT)
The crew from the San Diego Natural History Museum is hard at work at the ankylosaurid dinosaur excavation site. Photo courtesy of the San Diego Natural History Museum.

FIGURE 6.12 (ABOVE)
Ankylosaurid tooth still embedded in the enclosing rock matrix. Courtesy of the San Diego Natural History Museum; photo by the author.

the ditch and in so doing had broken up and removed the front portion of the skeleton. Even the construction workers pitched in to help search for any remaining pieces.

Following the removal of the ankylosaur from the site, Riney, Cerutti, Margo Rausch, and Aletha Patchen had the long, arduous job of chipping away the enclosing rock to expose the skeleton. Some of the preparation was also done by docent Cynthia Curlee, who over a two-year period worked along with others in public view in "The Bubble" at the San Diego Natural History Museum. A hole was cut in the glass so people could ask questions of the preparators.

The ankylosaur bones were solidly cemented in a siltstone concretion. After preparation, it became apparent that nearly the entire hindquarters of the animal had been retrieved, and both back legs were still articulated to the pelvis. Deméré (1988) described the

FIGURE 6.13

"The Bubble" at the San Diego Natural History Museum on June 4, 1987. Here Cynthia Curlee and Bradford Riney carefully work ankylosaur bones from their rocky matrix as museum visitors look on. Photo by Don Bartletti; courtesy of the *Los Angeles Times.*

FIGURE 6.14

Inside "The Bubble" at the San Diego Natural History Museum Bradford Riney carefully works with hammer and chisel on an ankylosaur leg. Photo by Don Bartletti; courtesy of the *Los Angeles Times.*

skeleton as "lying on its back with the legs splayed out to its sides, like some Cretaceous 'road kill.'" The bones had been lying on the muddy sea bottom, becoming a small reef for other organisms. Attached to the bones were oysters and snails, and in the muddy matrix was a shark tooth, suggesting that sharks may have scavenged on the carcass. Not only did the scientists find bones, but some dermal armor was also still in place. The armor patches were arranged like hexagonal floor tiles, and some of the loose scutes were domed or had a ridge running along their length. The ridged scutes may have been arranged in pairs running along the sides of the dinosaur's tail. Other loose bones and armor were found

FIGURE 6.15

"The skeleton had been lying on its back with the legs splayed out to its sides, like some Cretaceous 'road kill.'" From Coombs and Deméré 1996. Courtesy of the Paleontological Society.

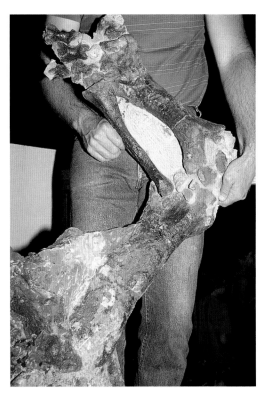

FIGURE 6.16 (LEFT)

Ankylosaur pelvic armor patches arranged like hexagonal floor tiles. Photo courtesy of the San Diego Natural History Museum.

FIGURE 6.17 (ABOVE)

One of the legs of the ankylosaur. Photo courtesy of the San Diego Natural History Museum.

scattered about in the matrix as well. The ankylosaur is one of the most complete dinosaur specimens ever found in California and is on display in the San Diego Natural History Museum.

On September 17, 1987, Riney found more dinosaur remains in the Cretaceous Point Loma Formation of San Diego—a metapodial (middle foot bone) fragment of a dinosaur. In the late 1980s, too, he found the proximal end of a carapace fragment from a fossil turtle in a conglomerate from the Late Cretaceous Cabrillo Formation. Unfortunately, this specimen has been lost.

BRADFORD RINEY (1954–)

Bradford Riney is the most successful dinosaur hunter in the state of California. In 1967, at the age of thirteen, he found his first dinosaur fossil in a sea cave in La Jolla, a single hadrosaur neck vertebra. Riney made a similar find again in 1976, when he found two vertebrae from a mosasaur (giant seagoing lizard) at Point Loma. His skills for finding important fossils were recognized by Tom Deméré of the San Diego Natural History Museum, who hired Riney as a curatorial assistant and field paleontologist.

The 1980s were very productive years for dinosaur discovery in San Diego County. In 1983 Riney found a hadrosaur femur and three years later thirteen hadrosaur vertebrae, all in

FIGURE 6.18
Bradford Riney in 1998 in a sea cave, where he made one of his early dinosaur discoveries. Photo by the author.

FIGURE 6.19
Robert D. Hansen at the site of his 1992 hadrosaur discovery. Photo courtesy of Robert D. Hansen.

the Carlsbad area. In 1987 he found a small foot bone of a dinosaur, and at a construction site in Carlsbad he made his most exciting find yet: an ankylosaur, one of the armored dinosaurs, a dinosaur never seen before in Alta California. It turned out to be one of the most complete dinosaur specimens ever found in the state: it even retained some of its dermal armor. Today it is on display in the San Diego Natural History Museum.

Riney is not only a talented paleontologist and geologist but also an excellent preparator. He has prepared most of the specimens that he found, as well as many others at the San Diego Natural History Museum. In addition, Riney is a talented artist and does paleontological illustrations at the museum, some for publication.

While conducting the research for this book I had the privilege of meeting Brad, and he took me to some of the sites of his dinosaur discoveries. He radiates his love of fossil hunting. Today he continues to search excavations, walk the beaches, and explore the cliffs of San Diego County looking for Mesozoic remains. I am sure that with his fantastic knack for fossil discovery he will continue to add to the wealth of specimens in the state.

Hansen, Calvano, and Peck: The 1990s

In 1992 Robert D. Hansen found the distal tibia of a hadrosaur in the Late Cretaceous Ladd Formation of Orange County. The life story of this Vietnam veteran is a familiar one. On returning from the war, he found himself unable to adjust to normal life: his marriage failed and his life became a nightmare of drugs, alcohol, living on the street, and occasional stints in jail. After years of hell, he finally got his act together, and while in recovery he used jewelry-making and rock, gem, and fossil hunting as a way to take his mind off things. It was while he was out looking for ammonites that he chanced upon the hadrosaur remains. Today Hansen has been free from drugs and alcohol for over two decades. His dinosaur bone has been donated to the Orange County Natural History Association, where it is on display at Ralph B. Clark Park, in Buena Park.

When he was a boy, Gino Calvano saw the fossil collection at Clark Park and became interested in paleontology. After receiving his B.A. in anthropology at UC Irvine in 1995, Calvano worked under the direction of Mark Roeder and David Whistler, curator for the Natural History Museum of Los Angeles County, as a paleontological monitor on the Eastern Transportation Corridor (a privately funded toll highway along the western base of the Santa Ana Mountains in Orange County). Here in 1996 he found foot bones, two cervical vertebrae, and a phalanx (toe) of a hadrosaurian dinosaur in the Late Cretaceous Williams Formation. The remains were prepared by California State University at Los Angeles graduate Gary Takeuchi. Calvano made another important discovery in 1996 when he found fossil turtle (*Basilemys*) remains, this time in the Late Cretaceous Ladd Formation of Orange

FIGURE 6.20
Gino Calvano (left) and
Mark Roeder display por-
tions of their hadrosaur
finds from Orange County.
Photo by the author.

FIGURE 6.21
Cycad frond on display at
the San Diego Natural History Museum. Courtesy
of the San Diego Natural
History Museum; photo
by the author.

County. He found more turtle remains (costal bones) in the same formation the next year.

Phil Peck, another monitor working for Roeder, also found some large dinosaur bones in Orange County's Williams Formation, but they are as yet unprepared, and hence remain unidentified.

Occasionally evidence of the Late Cretaceous foliage of an area comes to light through fossils. Plants often wash down streams or rivers into the sea, where they become waterlogged and sink to the bottom. Here sediment may cover the remains and preserve impressions of the plants. The San Diego Natural History Museum has a beautiful fossil cycad (sego palm) on display that was discovered in the Late Cretaceous marine rocks of San Diego County.

Orange and San Diego Counties have yielded a wealth of both dinosaurian and marine reptilian remains. As construction in the area continues and the sea persistently erodes into the coastal cliffs, more discoveries will be made and our knowledge of the Late Cretaceous paleoenvironment of southern California will grow.

PENINSULAR RANGES PROVINCE: BAJA DEL NORTE

Baja del Norte is the northern province of Baja California, Mexico. Here, about 160 miles south of San Diego, copious amounts of Mesozoic reptilian fossils have been collected. These fossils have contributed greatly to our understanding of the Late Mesozoic terrestrial life of the Peninsular Ranges of Baja as well as Alta California. (Before we go further, I wish to warn anyone even thinking of poaching fossils in Mexico to think again. Not only is it *illegal to collect,* but the consequences may be extreme.)

With the exception of Jurassic footprints from the Mojave and fragmentary remains from the Jurassic Trail Formation of the northern Sierra Nevada, the meager terrestrial deposits in Alta California have yielded few reptilian remains. There, almost all terrestrial information must be gleaned from materials that were washed into the sea. As a consequence, we must look to Baja California for the details we have not yet found in the state of California.

The Late Cretaceous geography of western Baja was probably not unlike today's, except that the peninsula had not yet become a peninsula but was still part of mainland Mexico. Unlike most of California, coastal northern Baja does contain Late Cretaceous terrestrial deposits; yet even here, these deposits are not abundant, for they exist in a restricted swath only some fifteen miles wide. According to William Morris (pers. comm. 1999), who led several paleontological expeditions in search of dinosaurs in Baja, at the eastern edge of these deposits are massive conglomerates (coarse gravel deposits), which originated from Late Cretaceous torrential flooding coming out of a mountainous interior. Westward these gravels typically grade into deposits laid down in braided channels and deltas near the mouths of Cretaceous rivers. Some of these deposits interlap with shallow marine deposits. On two occasions, dinosaur remains were found in association with marine fossils (Morris 1974b), and during the excavation of one of these a hadrosaurian tail section was found to be lying on top of an ammonite (Morris, pers. comm. 1999).

There is evidence of periods of extreme dry weather. One hadrosaurian dinosaur found in this area appears to have desiccated after it died, then was buried in sediment. Molds of the dried skin are preserved, as are the egg cases of primitive botflies that were probably feeding on the carcass (Morris 1974b).

These Baja outcrops have yielded a wealth of information about the paleofauna of the

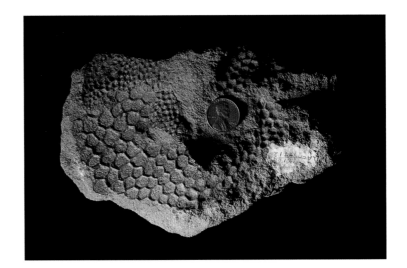

FIGURE 6.22
Molds of hadrosaur (*Lambeosaurus*) skin with scale patterns. Courtesy of the Vertebrate Paleontology Department, Natural History Museum of Los Angeles County; photo by the author.

Peninsular Ranges Province as well. In addition to animal fossils, an important flora has been discovered, allowing us to infer much about the habitat in which the animals lived. Along with the remains of silicified conifer and palm logs, the leaves of ginkgo and other deciduous trees have been found. Some of the trees had occasional root branches still intact, indicating that although they were transported under high-energy conditions, they did not travel far (Morris 1974b). Some of the conifers, which include *Araucaria* (monkey puzzle tree) and redwoods (*Sequoia*), were found in life position with their roots still penetrating the sandstones below (Morris 1974b). Vines have been found as well. Many of these plants would have provided food for the herbivorous dinosaurs and other animals of the area.

Today the west coast of northern Baja is a coastal desert. Though generally dry, it is favored by summer morning fogs. The fogs, which give fossil enthusiasts some respite from the heat, can render the landscape beautiful and serene. Hence we have fossil sites with the names of Misty Hill 1 and Misty Hill 2, where dinosaurs and crocodilians were found. Once the fogs burn off, the summer days may be very hot, not unlike those in the San Diego area; nevertheless, the temperatures can be variable, and the precipitation as well, with the remains of hurricanes, or "tabascos" as they are locally called, being not uncommon. Winter storms occasionally penetrate south and provide moisture here as well.

Most of the time, however, the area experiences perpetual drought, so although plant life does not have it easy, conditions are ideal for fossil collecting: there is lots of unvegetated ground to scour. In addition, when the infrequent rains do come, they are often torrential, causing erosion; if loose sedimentary rocks are present, badlands conditions are the inevitable

FIGURE 6.23 (LEFT)
Fossil tree trunk projecting
from a sandstone. Photo
from the collection of
Douglas Macdonald.

FIGURE 6.24 (BELOW)
Badlands yielding dinosaur
bones in Baja. Photo from
the collection of Douglas
Macdonald.

result. While badlands are usually extremely rough terrain with steep, treacherous slopes, they are prime areas in which to find fossils.

I have led field trips to Baja and found it to be no easy task, especially for an extended stay. At least in the badlands of Montana or South Dakota, there is no language barrier. Gasoline, car parts, and groceries are not far away, and you can find a mechanic and even a hospital if you need one. None of this is true for much of Baja California.

When fieldwork began in the region, the highway was not paved. William Morris (1966a [1973]), then a research associate with the Natural History Museum of Los Angeles County

(LACM) and associate professor of geology at Occidental College, describes working in Baja during the 1960s: "The areas under investigation are isolated, and the poorly maintained roads continue to pose difficulties in collecting, in gaining access into likely looking territory, and in the mechanical maintenance of vehicles. Mules, burros, and heavy-duty trucks have been utilized, but in most areas backpacking of supplies and equipment is necessary. Traveling the interior of Baja California is still an extreme trial for the durability of both vehicles and personnel. Common failures are broken springs, flat tires, and breakdown of electrical systems."

Mexico is increasingly taking its place in the science of paleontology. From the early 1950s to the mid-1970s the Mexican government allowed U.S. scientists with the proper permits to collect in northern Baja. At that time, under mutual agreement, almost all of these Mesozoic reptilian remains were curated at LACM. Inevitably, some contention as to who owned the fossils arose. As a result, most of these fossils have been rightfully returned to Mexico, but at least for a time American scientists were able to study and describe some of them.

Early Discoveries: 1953 to 1960

According to Morris (1965), the first hint of possible dinosaur remains in Baja California came from an unidentifiable bone scrap found in 1925 in Cretaceous badlands about 160 miles south of the Mexican border. The first recognizable dinosaur remains were discovered in 1953 by J. Wyatt Durham and Joseph H. Peck Jr. of the University of California, Berkeley, while searching for fossils in the El Gallo Formation. Peck's field notes for June 21, 1953, mention "fragments of dinosaur from a small draw cut in a yellow sand." These turned out to be foot bones from two small hadrosaurs.

The UCMP collection has one entry of dinosaur bone fragments found by Frank H. Kilmer in 1957. The tags identifying stored fossils further indicate that from December 30, 1959, to January 9, 1960, Kilmer and Russell led a paleontological expedition to Baja. This was a California Academy of Sciences expedition, supported by the Belvedere Scientific Fund, which in turn was sponsored by Kenneth K. Bechtel of the Bechtel Corporation, an international construction firm. Bechtel's interest in Baja was sparked by his trips to the peninsula with writer Joseph Wood Krutch and botanist Ira Wiggins (Krutch 1961). In the Berkeley collections there are five entries that match these dates: two for dinosaur bone fragments collected by Kilmer, and three other unidentified fragments that do not list the name of a collector.

E. H. Colbert, then curator of reptiles at the American Museum of Natural History, collected several specimens for Berkeley on March 9, 1960. At two sites he found theropod teeth, the first theropod remains found in Baja. He found other miscellaneous dinosaur teeth from other sites and at yet another location a hadrosaur jaw fragment with teeth. He also collected a hadrosaur vertebra centrum and the distal end of a hadrosaur femur.

FIGURE 6.25
The first recognizable dinosaur remains found in Baja:
foot and toe bones from a
small hadrosaur. Courtesy of
the University of California
Museum of Paleontology;
photo by the author.

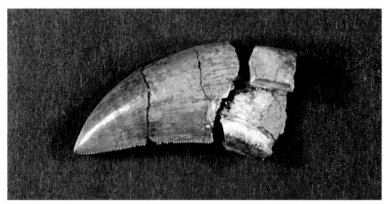

FIGURE 6.26
Theropod tooth from Baja.
Note serrated edges, perfect
for ripping flesh. Courtesy
of the Vertebrate Paleon-
tology Department, Natural
History Museum of Los
Angeles County; photo by
the author.

Morris, Garbani, and Crew: 1960s

In 1964 Shelton P. Applegate and Harley Garbani, both working with the Natural History Museum of Los Angeles County and Occidental College, made fossil discoveries in Baja. These led to another expedition that same year organized by LACM and Occidental College, during which William Morris found the partial skeleton of a dinosaur in the Cretaceous badlands (Morris 1974b). These discoveries became the catalyst for a major project—the first expedition for the specific purpose of finding dinosaurs in Baja—initiated by Morris and sponsored by the National Geographic Society (NGS) in the summer field season of 1965 (Morris 1968c).

Morris was the research director for the project, with support from both LACM and the Universidad Autónoma de Baja California. His senior field assistant was James (Harley)

FIGURE 6.27
William Morris (left) and
field crew members using
gasoline-powered tools to
excavate dinosaur bones in
the badlands of Baja in
1966. Courtesy of the
Vertebrate Paleontology
Department, Natural
History Museum of Los
Angeles County.

Garbani, preparator for LACM. James MacDonald, senior curator of vertebrate paleontology at LACM, also participated (Morris 1965).

A second expedition was mounted from June 17 to September 2, 1966, again directed by Morris and with Garbani as chief field assistant. This foray was sponsored by Universidad Autónoma de Baja California–Escuela Superior de Ciencias Marinas, NGS, and LACM. Field assistants were Adolfo Molina, Federico Zihel Helbling, Luis Gustavo Alvarez, Luis José Terui Trujano, Eric Austin, John Claque, Richard Clark, Robert Johnson, Richard LeFever, Andrew Steben, and William Seekins. Morris was fortunate to have such a large crew: excavating dinosaur bones in Baja is no easy task. Because of the remoteness of many of the discoveries, large dinosaur bones had to be broken, encased in burlap and plaster, and then hand-carried for miles over terrain that even a burro could not traverse; self-contained gasoline-powered impact hammers and a rock saw were helpful (Morris 1965). The summer of 1966 yielded the partial remains of hadrosaurs, plus compressed doubly serrate teeth from carnivorous dinosaurs similar to *Gorgosaurus* (Morris 1967c).

The 1967 summer field season, which ran from June 19 to September 2, was again sponsored by NGS and LACM, with Morris and Garbani continuing in their previous roles. Richard LeFever and David Lowe came along as junior assistants for the first half of the season; they were joined by Pedro Fonseca, a resident of El Rosario (who stayed the entire season), and Mario Torres from the University of Baja California. Fonseca, says Morris (pers. comm. 1999), was

FIGURE 6.28
Pedro Fonseca and his children on laundry day in 1967.
Photo from the collection of Douglas Macdonald.

"the strongest skinny man in the world. He found us working in his 'back yard' and seeing a section of the neck we were excavating said, 'Ah, espina' [Ah, the spine]." They hired him on the spot and he very capably managed students and local citizens on the job. Douglas Macdonald described Fonseca (field notes of August 15, 1967) as "an undernourished Gregory Peck—good sense of humor—hard worker—tireless walker—sharp eyed—stronger than he looks."

In the second half of the season the junior assistants were Alan Tabrum, Kent Valmassy, and Frank Bain. Douglas Macdonald, president of the museum alliance, joined the crew for a week, and NGS photographer Michael Hoover was with them for two.

Among the equipment this time, in addition to a Dodge Powerwagon and an International Travelall, were a small tractor, a rock saw, and two jackhammers—and it all came in handy, for that summer they discovered part of an extremely large hadrosaur (*Lambeosaurus*), estimated at fifty-four feet when alive (Morris 1972). Morris describes one find as follows: "The pelvis, including sacrum, together with tibia, fibula, and several dorsal vertebrate [*sic*] were collected. Vertebrate [*sic*] appear to continue into the hillside however, the nature of the matrix, and the amount of overburden, made it impossible to collect the entire fossil in the amount of time remaining. Future work would necessitate blasting in order to ascertain how complete the specimen is and to collect the portions present."

A few days' worth of excerpts from the field notes of Douglas Macdonald give a good flavor of what that expedition was like:

FIGURE 6.29 (LEFT)
William Morris uses a gas-powered drill to excavate a dinosaur tibia. Photo from the collection of Douglas Macdonald.

FIGURE 6.30 (RIGHT)
Richard Clark, Eric Austin, and Andrew Steven removing a block containing dinosaur vertebra. Courtesy of the Vertebrate Paleontology Department, Natural History Museum of Los Angeles County.

Tuesday, Aug. 15th Up at 0600—breakfast—load gear—off to site. Walk into badlands, then up arroyo to site of large bone.—Badlands full of petrified wood—take pictures—narrow walled, sandy-bedded arroyos—gray and reddish tilted beds overlain and interspersed with sandstone layers.—Large plaster block has to go out on a crude 2 × 4 litter—down canyon—dry waterfall—then to tractor-truck and over "killer hill" to jeep. We make it with two 200# blocks and one small 100-pounder. . . .

Thursday, Aug. 17th . . . back to embedded tibia [lower hind leg bone]. Alan Tabrum and I prospect all morning—much wood—no bone. Lunch at the tibia, then pack out last of smaller blocks. The tibia is in solid rock, and a devil to remove. While prospecting I find a 10- or 12-foot [petrified] log, slowly exfoliating into tiny pieces. Also find a superb specimen of what may be a conifer—complete with bark on it, and set it by the trail for later pick-up. . . . Back over KH [killer hill] with 100# block—almost at a run. Back in camp we find that we have been visited by coyotes in our absence. Our camp is on a wide plain with a large sand dune between us and the sea. At high tide we can hear the waves rolling in like distant thunder. They crash down and then retreat with a great rattling of beach rocks. . . .

Friday, Aug. 18th Doc Morris and Mike Hammer head for L.A. with blocks collected so far. (Mike was a photographer sent by Nat'l Geo to film the expedition. He was a tall, quiet, very athletic sort whose idea of cooling off at a day's end was to swim out into very rough ocean until he was out of sight, then eventually return refreshed, wondering why we were so worried—.) Jim, Pedro, Al and I head off south through the old town to collect parts of a carnosaur. We find a few scraps and some teeth, and the first arrowhead I have ever found. . . .

FIGURE 6.31 (LEFT)
Pedro Fonseca (above) and
another crew member ped-
estal the tibia before it is
wrapped in burlap and plaster for transport. Photo from
the collection of Douglas
Macdonald.

FIGURE 6.32 (RIGHT)
Another dinosaur bone en
route to "Killer Hill." Photo
from the collection of
Douglas Macdonald.

Tuesday, Aug. 22nd Up to the original site of the tibia. We remove it in pieces, and uncover more bone using the gasoline-driven jackhammer and cold chisels. After lunch, more prospecting as this is the last day in the field. Still slim pickings. Work our way down the coast to beautiful beaches, sea cave, etc. Once more over KH [killer hill] with pack full of dinosaur bone. One last ride over the "torture-bahn" to camp. Dr. Morris due in at 1900, but doesn't show up (?).

The arduous tasks of finding fossils and excavating continued, and the crew was also lucky enough to find further evidence of carnosaurs that season. Their luck ended two weeks sooner than expected, however, as the following narrative by Macdonald explains:

[We] wound up the season in an unusual and totally unexpected fashion. Hurricane KATRINKA, which made headlines by devastating San Felipe on the East Coast of Baja California, also caused a flash flood in El Rosario, on the West Coast, that destroyed homes, farmes [*sic*], and livestock as well as taking the lives of several residents of this little town.

Our field camp was located on a wide plain, west of town, and close to the beach. It was sheltered from the sea by a continuous bar of sand and wave-worn boulders, rising ten to twelve feet above the plain and completely closing the mouth of the valley. Following KATRINKA's progress from radio news broadcasts, and alerted by worsening weather conditions in the immediate area, Dr. Morris made an important decision. He ordered all specimens, machinery, and tools removed from the digging sites and had them loaded aboard the party's vehicles. Late in the night, gale-force winds and heavy rains drove the party to

WILLIAM J. MORRIS (1923–2000)

William Morris is responsible for more dinosaur discoveries on the west coast of North America than any other person. He and his paleontological teams in Baja California are credited with the discovery of new dinosaurs, a new bird, and new crocodilians, as well as several fossil lizards and turtles.

Born in Baltimore in 1923, Morris served in the U.S. Army from 1942 to 1945. He earned a B.A. from Syracuse University in 1947, and an M.A. (1948) and Ph.D. (1950) from Princeton. His first teaching job was as assistant professor of geology at Texas A & M, and he continued there as an associate professor until 1955. He then took a position at Occidental College, where he became a full professor of geology. In 1957 he began a

FIGURE 6.33

This 1965 photo shows William Morris (right) and Jim (Harley) Garbani inspecting a hip bone (left ischium) from the duck-billed dinosaur *Lambeosaurus laticaudus* discovered in Baja California. Courtesy of the Vertebrate Paleontology Department, Natural History Museum of Los Angeles County.

high ground. The rest of the night was spent in towing stranded vacationers to safety with the stoutest of the four-wheel-drive trucks.

A gray and dismal dawn revealed the abandoned tents to be down and in tatters. After salvaging what was still usable, Dr. Morris decided to waste no further time in starting the party back to Los Angeles. In the light of what was to follow, his correct estimate of the situation and his prompt decision saved the entire summer's work and the major equipment of his field party, and quite possibly avoided injury or even loss of life among its members.

The evening of this same day, a flash-flood, born of the previous night's torrential rains, swept down from the mountains to the east of El Rosario, carrying away everything in its path. Until it was breached, our sheltering bar dammed a lake five to six feet deep behind it. Then it failed and the flood burst out into the sea, gnawing a hundred-foot wide hole near our former camp site, and totally removing a 500 yard long section of the bar, at the south end of the beach near the cliffs. Today waves break unopposed into what was once a quiet brackish pond—now a mile long estuary, and the people of El Rosario are trying to put their lives back together.

Upon his return, wishing to express the Museum's concern and regard for the people of this stricken village, Dr. Morris explained what had happened and called on the staff of the LACMNH for assistance. The response was immediate and open hearted. The party's two trucks were loaded with gifts of clothing and tools, but still more remained. Mr. Walter Laird, a member of the Museum Alliance, volunteered his plane and piloting skills to help long association with the Natural History Museum of Los Angeles County, where he was a research associate in vertebrate paleontology.

According to Douglas Macdonald, Morris was the driving force behind the arduous task of doing vertebrate paleontology in Baja. He not only raised the money through the National Geographic Society and other organizations, he also made arrangements for all the equipment and a photographer. Morris was honored with the Arnold Guyot Award from the National Geographic Society in 1968 and was a fellow of the Geological Society of America. He served as director of the Southern California Academy of Sciences from 1958 to 1970. Morris wrote many papers on his paleontological work and discoveries of Mesozoic reptiles in both Alta and Baja California. His contribution to our knowledge of West Coast Mesozoic vertebrates and the Late Cretaceous paleoenvironment here is enormous.

move more of the clothing, and Museum Alliance funds underwrote the actual cost of the flight. In this way the maximum amount of supplies were carried to El Rosario while the need was the greatest. The two trucks were driven non-stop to El Rosario by Dr. Morris and Dr. J. R. MacDonald, Senior Curator of Vertebrate Paleontology of our Museum. Upon arrival, all supplies were turned over to Sr. Jesus P. Pirado, the Delegado of El Rosario, for distribution to the people of the town. Dr. Morris will return to El Rosario, to continue his dinosaur-hunting. May he find that his thoughtful "mission" has contributed in part to the prompt and complete recovery of this friendly village.

Morris explains in his notes that his previous experience with storms in Baja had made him wary. In this storm he estimated the wind speed at between eighty and one hundred miles per hour. Although Morris moved all his people and most of the heavy equipment to high ground, the erosion and deposition from the heavy rain and high winds ended up burying many personal items as well as tents, cots, and sleeping bags.

The 1968 season shows two entries in the LACM collection for one hadrosaur. This could be from the hadrosaur they had started to uncover in 1967 just before the hurricane caused their early departure. The field season of 1969, funded in part by a National Science Foundation grant, was spent in reconnaissance, looking for new sites and fossils that they might excavate at a later date.

Morris and Crew: 1970s

In the summer of 1970 the LACM/Occidental group again went to northern Baja prepared to collect. This time Morris was joined by Jason A. Lillegraven, a research associate at LACM and associate professor of zoology at California State College, San Diego, and Ismael

Ferrusquia-Villafranca, a paleontologist at the Instituto de Geológica, Universidad Nacional Autónoma de México. Field assistants were Pedro Fonseca, Alan Tabrum, Bruce Burns, Richard Berggren, and Steven Poppadakis. This expedition was sponsored by LACM and the Instituto de Geológica, with NGS providing financial backing. The field vehicles were an International Carryall and a Jeep Universal and trailer.

On this trip, Morris's group concentrated on locating reptilian remains, while Lillegraven's group searched primarily for sediment they might screen for Cretaceous mammalian remains—a tedious job at best. Ferrusquia-Villafranca assisted both parties.

Torrential rains in 1969 had exposed many new fossils. James (Harley) Garbani discovered the remains—several bones, as well as part of the skull and jaw complete with teeth—of a new, large theropod dinosaur, *Labocania anomala.* These carnivorous remains were found in association with hadrosaur ribs in the Late Cretaceous La Bocana Roja Formation.

The next July, 1971, Garbani discovered a single dermal scute of an ankylosaur: the first evidence of an ankylosaur west of the Rocky Mountains.

In the spring of 1973 Michael Greenwald of LACM came across some small bones in the museum's Late Cretaceous collection from northern Baja that had been collected two years earlier by Garbani (Garbani, pers. comm. 2002). They turned out to be from the first terrestrial bird discovered since *Archaeopteryx.* Because of this important discovery a proposal was submitted to NGS to go to Baja in search of Cretaceous birds. NGS approved the expedition, which was subsequently mounted, under the direction of LACM and the Instituto de Geológica, between August 25 and September 17, 1973. The project directors were Morris and Ferrusquia-Villafranca, with Garbani acting as field party leader. Field assistants included Michael Greenwald and Erich Lichtward of LACM, Oscar Carranza of the Instituto de Geológica, and Pedro Fonseca.

The group succeeded in finding more Mesozoic bird material, as well as "the most complete and undistorted specimen of a Cretaceous crocodilian ever to have been reported" (Morris, pers. comm., 1999). Never before described, this specimen had characteristics of both alligators and crocodiles and may be an early relative of the modern alligators. Greenwald found probable dinosaur eggshell fragments and early mammalian material on this expedition as well.

In 1974 NGS sponsored three further expeditions co-led by LACM and the Instituto de Geológica for the purpose of paleontological reconnaissance of the length of the Baja Peninsula. A newly paved highway running the length of Baja provided an excellent opportunity for extensive exploration. Only a small part of this reconnaissance, however, was dedicated to the Late Cretaceous dinosaur-bearing strata (Morris 1974a). To date, this was apparently the last expedition to Baja in search of dinosaur remains.

Many bones of smaller reptiles have been found in Baja, all of which were sent to Richard

FIGURE 6.34
Dinosaur excavation can be precarious in the badlands of Baja. Courtesy
of the Vertebrate Paleontology Department, Natural History Museum
of Los Angeles County.

Estes of San Diego State University for study (Morris 1968c, 1974b). Estes identified a teid
lizard (similar to the present-day *Paraglyphanodon*) and a small crushing-toothed lizard of
unknown affinities. Among turtles he mentions a pustulse turtle, which Hutchison (pers.
comm. 2000) says is in the genus *Naomichelys*—a genus known only from fragments of
its shell.

Baja California has added immensely to our knowledge of the Late Cretaceous fauna and
flora of peninsular California, but as is evident, paleontological work here has lapsed in recent years. I have tried to find out whether there have been any other expeditions to find
Mesozoic reptiles in Baja since the mid-1970s, but have heard of none. Perhaps paleontologists from Mexico and the United States will again team up one day. The scientific benefits
could be very exciting.

With only one hundred years of effort, the collection of Mesozoic reptilian fossils in California has just begun. In just the five years that saw the research and writing of this book, several plesiosaur remains, a couple of mosasaur remains, four more dinosaurs, new turtles, the first Mesozoic birds, and the first pterosaur remains have been found in California. It is my hope that this book will spur interest that will lead to many more such finds. Old untold discoveries may come to light and even more history unearthed. As time passes this book will become but a small chapter in a large and varied encyclopedia of Mesozoic reptilian remains yet to be discovered.

APPENDIX

Summary of the Mesozoic Reptilian Fossils of California

REPOSITORY SYMBOLS

AMNH: American Museum of Natural History

BC: Butte College

CAS: California Academy of Sciences

CIT: California Institute of Technology

CSUF: California State University Fresno

ETC: Eastern Transportation Corridor (Orange Co.)

GÖHRE: Eric Göhre

LACM: Natural History Museum of Los Angeles County

MPC: Monterey Peninsula College

OCNHF: Orange County Natural History Foundation

OCPC: Orange County Paleontological Collection

PU: Princeton University

SBCM: San Bernardino County Museum

SC: Sierra College Natural History Museum

SDNHM: San Diego Natural History Museum

UCMP: University of California Museum of Paleontology

YPM: Yale-Peabody Museum of Vertebrate Paleontology

FOSSIL REPTILES FROM BAJA CALIFORNIA

Baja dinosaurs

GROUP	TYPE	GENUS	SPECIES	REPOSITORY	SPECIMEN NO.	ELEMENTS	AGE	FORMATION	FINDER	FIND DATE	PROVINCE
Dinosauria				LACM to Mex.	42686/HJG 699	Eggshell frag.	Late Cret.	El Gallo	Greenwald, M. T.	1973	Baja
Dinosauria				LACM to Mex.	42688/HJG 700	Eggshell frag.	Late Cret.	El Gallo	Greenwald, M. T.	1973	Baja
Dinosauria				LACM to Mex.	20873/HJG 62	Vertebrae caudal	Late Cret.	La Bocana Roja	Morris, W. J.	1967	Baja
Dinosauria				LACM to Mex.	42717/HJG 705	Eggshell frag.	Late Cret.	El Gallo	Garbani, H. J.	1973	Baja
Dinosauria				LACM to Mex.	42671/HJG 695	Bone	Late Cret.	El Gallo	Garbani, H. J.	1973	Baja
Dinosauria				UCMP		Lg. bone frags. + vertebra centrum frag.	Late Cret.	El Gallo	Kilmer, F. H.	1959–60	Baja
Dinosauria				UCMP		Bone frags.	Late Cret.	El Gallo	UCMP	1959	Baja
Dinosauria				UCMP		Bone frags.	Late Cret.	El Gallo	UCMP	1959	Baja
Dinosauria				UCMP		Bone frags.	Late Cret.	El Gallo	Kilmer, F. H.	1957	Baja
Dinosauria				UCMP		Misc. teeth	Late Cret.	El Gallo	Colbert, E. H.	1960	Baja
Dinosauria				UCMP		Misc. teeth	Late Cret.	El Gallo	Colbert, E. H.	1960	Baja
Dinosauria				UCMP		Bone frags.	Late Cret.	El Gallo	Kilmer, F. H.	1959	Baja
Dinosauria				UCMP	54994	Bone frags.	Late Cret.	El Gallo			Baja
Dinosauria				UCMP		Bone frags.	Late Cret.	El Gallo			Baja
Dinosauria				UCMP		Bone frags.	Late Cret.	El Gallo	UCMP		Baja
Dinosauria				UCMP		Bone frags.	Late Cret.	El Gallo	UCMP		Baja
Dinosauria				UCMP		Bone frags.	Late Cret.	El Gallo	UCMP		Baja
Dinosauria				UCMP		Bone frags.	Late Cret.	El Gallo	UCMP		Baja
Dinosauria				UCMP		Bone frags.	Late Cret.	El Gallo	UCMP	1959–60	Baja
Ornithischia	Ankylosauria			LACM to Mex.	29000	Dermal plate	Late Cret.	El Gallo	Garbani, H. J.	1971	Baja
Ornithischia	Hadrosauridae			LACM to Mex.	17702	Vertebrae + ischia	Late Cret.	El Gallo	Morris, W. J.	1966	Baja
Ornithischia	Hadrosauridae			LACM	17702	Vertebrae	Late Cret.	El Gallo	Morris, W. J.	1966	Baja
Ornithischia	Hadrosauridae			UCMP	43251/?A-9525	Podials (2)	Late Cret.	El Gallo	Durham, W. J. & Peck, J. (or M. H. Oakes)	1953	Baja
Ornithischia	Hadrosauridae			LACM	20874	Ischium, proximal end	Late Cret.	El Gallo	Morris, W. J.		Baja
Ornithischia	Hadrosauridae			LACM to Mex.	17716	Humerus	Late Cret.	El Gallo	Morris, W. J.	1966	Baja
Ornithischia	Hadrosauridae			LACM	17716	Humerus	Late Cret.	El Gallo	Morris, W. J.	1966	Baja
Ornithischia	Hadrosauridae			LACM		Humerus	Late Cret.	El Gallo	Morris, W. J.	1970	Baja
Ornithischia	Hadrosauridae			LACM		Vertebra caudal	Late Cret.	El Gallo	Morris, W. J.	1966	Baja
Ornithischia	Hadrosauridae			UCMP	43251	Foot bones (2 ind.)	Late Cret.	Rosario			Baja
Ornithischia	?Hadrosauridae			UCMP		Vertebra centrum	Late Cret.	El Gallo	Colbert, E. H.	1960	Baja
Ornithischia	Hadrosauridae			UCMP		Femur, distal end	Late Cret.	El Gallo	Colbert, E. H.	1960	Baja
Ornithischia	Hadrosauridae			UCMP		Jaw frag. w/ teeth +	Late Cret.	El Gallo	Colbert, E. H.	1960	Baja
Ornithischia	Hadrosauridae			LACM to Mex.	23625	Skeleton (incompl.)	Late Cret.	La Bocana Roja	Morris, W. J.	1968	Baja

GROUP	TYPE	SPECIES	REPOSITORY	SPECIMEN NO.	ELEMENTS	AGE	FORMATION	FINDER	FIND DATE	PROVINCE
Ornithischia	Hadrosauridae		LACM	23625	Humerus, rt.	Late Cret.	La Bocana Roja	Morris, W. J.	1968	Baja
Ornithischia	Hadrosauridae		LACM to Mex.		Caudal vertebra	Late Cret.	El Gallo	Morris, W. J.	1966	Baja
Ornithischia	Hadrosauridae		LACM to Mex.	52463	Vertebra + ischium frags.	Late Cret.	El Gallo	Morris, W. J.		Baja
Ornithischia	Hadrosauridae		LACM to Mex.	52462	Vertebra frags.	Late Cret.	El Gallo	Morris, W. J.		Baja
Ornithischia	Hadrosauridae		LACM to Mex.	52461	Vertebra + limb bones	Late Cret.	El Gallo	Morris, W. J.		Baja
Ornithischia	Hadrosauridae		LACM to Mex.	52460	Mandible frag.	Late Cret.	El Gallo	Morris, W. J.		Baja
Ornithischia	Hadrosauridae		LACM to Mex.	29004	Tooth	Late Cret.	La Bocana Roja	Morris, W. J.		Baja
Ornithischia	Hadrosauridae		LACM to Mex.	28236	Humerus	Late Cret.	El Gallo	Morris, W. J.	1971	Baja
Ornithischia	Hadrosauridae		LACM to Mex.	28235	Humerus	Late Cret.	El Gallo	Morris, W. J.	1971	Baja
Ornithischia	Hadrosauridae		LACM to Mex.	28234	Humerus	Late Cret.	El Gallo	Morris, W. J.	1971	Baja
Ornithischia	Hadrosauridae		LACM to Mex.	17717/IGM 7-66-3-9	Fibula	Late Cret.	El Gallo	Morris, W. J.	1966	Baja
Ornithischia	Hadrosauridae		LACM to Mex.	20885/WJM 7-66-2-1	Fibula (incompl.)	Late Cret.	El Gallo	Morris, W. J.	1966	Baja
Ornithischia	Hadrosauridae		LACM to Mex.	20884/WJM 7-66-3-8	Dentary + pubis frags.	Late Cret.	El Gallo	Morris, W. J.	1966	Baja
Ornithischia	Hadrosauridae		LACM to Mex.	20883/WJM 7-66-1-13	Vertebra frags.	Late Cret.	El Gallo	Morris, W. J.	1966	Baja
Ornithischia	Hadrosauridae		LACM to Mex.	20875/JHG [HJG] 61	Innominate ischium, distal end	Late Cret.	La Bocana Roja	Morris, W. J.	1967	Baja
Ornithischia	Hadrosauridae		LACM to Mex.	17713/IGM 7-66-3-4	Dentary frag.	Late Cret.	El Gallo	Morris, W. J.	1966	Baja
Ornithischia	Hadrosauridae		LACM to Mex.	17710/IGM 7-66-3-4	Rib (incompl.)	Late Cret.	El Gallo	Morris, W. J.	1966	Baja
Ornithischia	Hadrosauridae		LACM to Mex.	17709/IGM 7-66-3-4	Vertebra thoracic	Late Cret.	El Gallo	Morris, W. J.	1966	Baja
Ornithischia	Hadrosauridae		LACM to Mex.	17708/IGM 7-66-3-4	Innominate ischium, distal end	Late Cret.	El Gallo	Morris, W. J.	1966	Baja
Ornithischia	Hadrosauridae		LACM to Mex.	17707/IGM 7-66-3-4	Humerus	Late Cret.	El Gallo	Morris, W. J.	1966	Baja
Ornithischia	Hadrosauridae		LACM to Mex.	17703/JHG [HJG] 61	Innominate ischium, distal end	Late Cret.	El Gallo	Morris, W. J.	1966	Baja
Ornithischia	Hadrosauridae		LACM to Mex.	17700/ WJM 6-66-1-6	Tooth	Late Cret.		Morris, W. J.	1966	Baja
Ornithischia	Hadrosauridae		LACM to Mex.	17699/ WJM 6-66-1-5	Tooth	Late Cret.		Morris, W. J.	1966	Baja
Ornithischia	Hadrosauridae		LACM to Mex.	17698/ WJM 6-66-1-4	Vertebra caudal	Late Cret.		Morris, W. J.	1966	Baja
Ornithischia	Hadrosauridae		LACM to Mex.	101172	Teeth	Late Cret.	El Gallo	Garbani, H. J.	1971	Baja
Ornithischia	Hadrosauridae		LACM to Mex.	101170	Teeth	Late Cret.	El Gallo	Garbani, H. J.	1971	Baja
Ornithischia	Hadrosauridae		LACM to Mex.	42716/HJG 705	Tooth	Late Cret.	El Gallo	Garbani, H. J.	1973	Baja
Ornithischia	Hadrosauridae		LACM to Mex.	42707/HJG 704	Tooth	Late Cret.	El Gallo	Garbani, H. J.	1973	Baja
Ornithischia	Hadrosauridae		LACM to Mex.	42706/HJG 704	Tooth	Late Cret.	El Gallo	Garbani, H. J.	1973	Baja
Ornithischia	Hadrosauridae		LACM to Mex.	42640/HJG 689	Tooth	Late Cret.	El Gallo	Garbani, H. J.	1973	Baja
Ornithischia	Hadrosauridae		LACM to Mex.	42632/HJG 688	Tooth	Late Cret.	El Gallo	Garbani, H. J.	1973	Baja
Ornithischia	Hadrosauridae		LACM to Mex.	42584/HJG 685	Pes phalanx ungal	Late Cret.	El Gallo	Garbani, H. J.	1973	Baja
Ornithischia	Hadrosauridae		LACM to Mex.	29002	Vertebra cervical	Late Cret.	El Gallo	Garbani, H. J.	1971	Baja
Ornithischia	Hadrosauridae		LACM to Mex.	28996	Tooth + limb frags.	Late Cret.	El Gallo	Garbani, H. J.	1971	Baja
Ornithischia	Hadrosauridae		LACM to Mex.	28991	Innominate frags.	Late Cret.	El Gallo	Garbani, H. J.	1971	Baja
Ornithischia	Hadrosauridae		LACM to Mex.	101174/28992	Phalanx ungal	Late Cret.	El Gallo	Greenwald, M. T.	1971	Baja

GROUP	TYPE	GENUS	SPECIES	REPOSITORY	SPECIMEN NO.	ELEMENTS	AGE	FORMATION	FINDER	FIND DATE	PROVINCE
Ornithischia	Hadrosauridae			LACM to Mex.	101171/28993	Teeth	Late Cret.	El Gallo	Greenwald, M. T.	1971	Baja
Ornithischia	Hadrosauridae			LACM to Mex.	57872	Teeth	Late Cret.	El Gallo	Greenwald, M. T.	1973	Baja
Ornithischia	Hadrosauridae			LACM to Mex.	42684/HJG 697	Tooth	Late Cret.	El Gallo	Greenwald, M. T.	1973	Baja
Ornithischia	Hadrosauridae			LACM to Mex.	42667/HJG 693	Tooth	Late Cret.	El Gallo	Greenwald, M. T.	1973	Baja
Ornithischia	Hadrosauridae			LACM to Mex.	42644/HJG 690	Tooth	Late Cret.	El Gallo	Greenwald, M. T.	1973	Baja
Ornithischia	Hadrosauridae			LACM to Mex.	42639/HJG 689	Tooth	Late Cret.	El Gallo	Greenwald, M. T.	1973	Baja
Ornithischia	Hadrosauridae			LACM to Mex.	42587/HJG 687	Pes phalanx	Late Cret.	El Gallo	Greenwald, M. T.	1973	Baja
Ornithischia	Hadrosauridae			LACM to Mex.	29003	Humerus + frags.	Late Cret.	La Bocana Roja	Tabrum, A.	1970	Baja
Ornithischia	Hadrosauridae			LACM to Mex.	26757	Humerus, rt. (incompl.)	Late Cret.	El Gallo	Tabrum, A.	1970	Baja
Ornithischia	Hadrosauridae	Lambeosaurus		LACM	17715	Maxilla/premaxilla, lft.+	Late Cret.	El Gallo	Morris, W. J.	1968	Baja
Ornithischia	Hadrosauridae	Lambeosaurus		LACM	17712	Skin impressions	Late Cret.	El Gallo			Baja
Ornithischia	Hadrosauridae	Lambeosaurus		LACM	17715	Femur + ischium frag.	Late Cret.	El Gallo			Baja
Ornithischia	Hadrosauridae	Lambeosaurus cf.		UCMP	137303	Skin impressions	Late Cret.	El Gallo	Morris, W. J.	1971	Baja
Ornithischia	Hadrosauridae	Lambeosaurus cf.		UCMP	137303	Skin impressions	Late Cret.	El Gallo	Greenwald, M. T.	1982	Baja
Ornithischia	Hadrosauridae	Lambeosaurus cf.		LACM to UCMP	LACM17712	Skin impressions	Late Cret.	El Gallo	Greenwald, M. T.	1985	Baja
Ornithischia	Hadrosauridae	Lambeosaurus Gs cf.		LACM	17716/IGM 7-66-3-7	Humerus	Late Cret.	El Gallo	Morris, W. J.	1966	Baja
Ornithischia	Hadrosauridae	Lambeosaurus	laticaudus Gs?	LACM to Mex.	20874/JHG [HJG] 59	Skeleton (incompl.)	Late Cret.	El Gallo	Morris, W. J.	1967	Baja
Ornithischia	Hadrosauridae	Lambeosaurus	laticaudus Gs?	LACM to Mex.	20876/JHG [HJG] 58	Tibia, lft.	Late Cret.	La Bocana Roja	Morris, W. J.	1967	Baja
Ornithischia	Hadrosauridae	Lambeosaurus Gs cf.	laticaudus Gs?	LACM to Mex.	17715/IGM 7-66-3-6	Skeleton (incompl.) +	Late Cret.	El Gallo	Morris, W. J.	1966	Baja
Ornithischia	Hadrosauridae	Lambeosaurus Gs cf.	laticaudus Gs?	LACM to Mex.	17712/IGM 7-66-3-3	Vertebra thoracic	Late Cret.	El Gallo	Morris, W. J.	1966	Baja
Ornithischia	Hadrosauridae	Lambeosaurus Gs cf.	laticaudus Gs?	LACM to Mex.	17711/IGM 7-66-3-2	Tibia	Late Cret.	El Gallo	Morris, W. J.	1966	Baja
Ornithischia	Hadrosauridae	Lambeosaurus Gs cf.	laticaudus Gs?	LACM to Mex.	17706/IGM 7-66-3-2	Tibia	Late Cret.	El Gallo	Morris, W. J.	1966	Baja
Ornithischia	Hadrosauridae	Lambeosaurus Gs cf.	laticaudus Gs?	LACM to Mex.	17705/IGM 7-66-2-2	Vertebra caudal	Late Cret.		Morris, W. J.	1966	Baja
Ornithischia	Hadrosauridae	Lambeosaurus Gs cf.	laticaudus Gs?	LACM to Mex.	17704/101181	Tibia	Late Cret.		Morris, W. J.	1966	Baja
Ornithischia	Hadrosauridae	Lambeosaurus Gs cf.	laticaudus Gs?	LACM to Mex.	17702	Vertebra caudal	Late Cret.		Morris, W. J.	1966	Baja
Ornithischia	Hadrosauridae	Lambeosaurus Gs cf.	laticaudus Gs?	LACM to Mex.	28990	Vertebra + innominate	Late Cret.	La Bocana Roja	Garbani, H. J.	1971	Baja
Ornithischia	Hadrosauridae			LACM to Mex.	101163	Teeth	Late Cret.	El Gallo	Garbani, H. J.	1971	Baja
Saurischia				LACM to Mex.	101182/20889	Vertebra caudal + frags.	Late Cret.		Morris, W. J.	1966	Baja
Saurischia				LACM to Mex.	101184/20879	Teeth	Late Cret.	El Gallo	Morris, W. J.	1967	Baja
Saurischia				LACM to Mex.	101183/20879	Tooth	Late Cret.	El Gallo	Morris, W. J.	1967	Baja
Saurischia				LACM to Mex.	101173/28994	Tooth	Late Cret.	El Gallo	Greenwald, M. T.	1971	Baja
Saurischia				LACM to Mex.	101164/28992	Manus phalanx	Late Cret.	El Gallo	Greenwald, M. T.	1971	Baja
Saurischia	Carnosauria			LACM to Mex.	52459	Phalanx	Late Cret.	El Gallo	Morris, W. J.		Baja
Saurischia	Carnosauria			LACM to Mex.	52458	Teeth	Late Cret.	El Gallo	Morris, W. J.		Baja
Saurischia	Carnosauria			LACM to Mex.	20889/WJM 6-66-1-2	Phalanx, distal end	Late Cret.		Morris, W. J.	1966	Baja
Saurischia	Carnosauria			LACM to Mex.	17714/IGM 7-66-3-5	Teeth	Late Cret.	El Gallo	Morris, W. J.	1966	Baja
Saurischia	Carnosauria			LACM to Mex.	20879	Tooth	Late Cret.	El Gallo	Morris, W. J.	1967	Baja
Saurischia	Carnosauria			LACM to Mex.	17701/IGM 7-66-3-5	Tooth	Late Cret.		Morris, W. J.	1966	Baja

GROUP	TYPE	GENUS	SPECIES	REPOSITORY	SPECIMEN NO.	ELEMENTS	AGE	FORMATION	FINDER	FIND DATE	PROVINCE
Saurischia	Carnosauria			LACM to Mex.	17697/ IGM 7-66-1-3	Tooth	Late Cret.		Morris, W. J.	1966	Baja
Saurischia	Carnosauria			LACM to Mex.	17696/ IGM 7-66-1-1	Tooth	Late Cret.		Morris, W. J.	1966	Baja
Saurischia	Carnosauria			LACM to Mex.	42574/HJG 681	Tooth	Late Cret.	El Gallo	Garbani, H. J.	1973	Baja
Saurischia	Carnosauria			LACM to Mex.	42705/HJG 704	Tooth	Late Cret.	El Gallo	Garbani, H. J.	1973	Baja
Saurischia	Carnosauria			LACM to Mex.	42638/HJG 689	Manus phalanx	Late Cret.	El Gallo	Garbani, H. J.	1973	Baja
Saurischia	Carnosauria			LACM to Mex.	42631/HJG 688	Tooth	Late Cret.	El Gallo	Garbani, H. J.	1973	Baja
Saurischia	Carnosauria			LACM to Mex.	42563/HJG 679	Tooth	Late Cret.	El Gallo	Garbani, H. J.	1973	Baja
Saurischia	Carnosauria			LACM to Mex.	28998	Tooth	Late Cret.	El Gallo	Garbani, H. J.	1971	Baja
Saurischia	Carnosauria			LACM to Mex.	28997	Teeth	Late Cret.	El Gallo	Garbani, H. J.	1971	Baja
Saurischia	Carnosauria			LACM to Mex.	28993	Teeth	Late Cret.	El Gallo	Garbani, H. J.	1971	Baja
Saurischia	Carnosauria			LACM to Mex.	42687/HJG 700	Tooth	Late Cret.	El Gallo	Greenwald, M. T.	1973	Baja
Saurischia	Carnosauria			LACM to Mex.	42669/HJG 694	Tooth	Late Cret.	El Gallo	Greenwald, M. T.	1973	Baja
Saurischia	Carnosauria			LACM to Mex.	42570/HJG 680	Tooth	Late Cret.	El Gallo	Greenwald, M. T.	1973	Baja
Saurischia	Carnosauria			LACM to Mex.	42704	Tooth	Late Cret.	El Gallo	Fonseca, P.	1973	Baja
Saurischia	Carnosauria			LACM to Mex.	42685/HJG 698	Tooth	Late Cret.	El Gallo	Fonseca, P.	1973	Baja
Saurischia	Carnosauria			LACM to Mex.	42564/HJG 679	Tooth	Late Cret.	El Gallo	Fonseca, P.	1973	Baja
Saurischia	Coelurosauria	Chirostenotes		LACM to Mex.	58009	"Tooth"	Late Cret.	El Gallo	Lillegraven, J. A.	1970	Baja
Saurischia	Coelurosauria	Chirostenotes Gs cf.		LACM to Mex.	42586/HJG 687	"Tooth"	Late Cret.	El Gallo	Garbani, H. J.	1973	Baja
Saurischia	Coelurosauria Infor cf.			LACM to Mex.	42636/HJG 689	Tooth	Late Cret.	El Gallo	Greenwald, M. T.	1973	Baja
Saurischia	Dromaeosauridae			LACM to Mex.	58010	Tooth	Late Cret.	El Gallo	Lillegraven, J. A.	1970	Baja
Saurischia	Saurornithoididae	Saurornithoides Gs cf.		LACM to Mex.	42585/HJG 687	Tooth	Late Cret.	El Gallo	Garbani, H. J.	1973	Baja
Saurischia	Saurornithoididae	Saurornithoides Gs cf.		LACM to Mex.	42637/HJG 689	Tooth	Late Cret.	El Gallo	Garbani, H. J.	1973	Baja
Saurischia	Saurornithoididae			LACM to Mex.	42675/HJG 696	Tooth	Late Cret.	El Gallo	Garbani, H. J.	1973	Baja
Saurischia	Theropoda	Labocania	anomala	LACM to Mex.	20877/JHG 65	Skull, dentary frag. +	Late Cret.	El Gallo	Garbani, J.? /Morris, W. J.	1970?/67	Baja
Saurischia	Theropoda			LACM to Mex.	42703/HJG 704	Phalanx ungal	Late Cret.	El Gallo	Garbani, H. J.	1973	Baja
Saurischia	Theropoda			LACM to Mex.	42571/HJG 680	Manus phalanx, distal end	Late Cret.	El Gallo	Garbani, H. J.	1973	Baja
Saurischia	Theropoda			LACM to Mex.	57871	Tooth frag.	Late Cret.	El Gallo	Greenwald, M. T.	1973	Baja
Saurischia	Theropoda			LACM to Mex.	42565/HJG 679	Manus phalanx ungal	Late Cret.	El Gallo	Fonseca, P.	1973	Baja
Saurischia	Theropoda			LACM	17704	Tooth	Late Cret.	El Gallo			Baja?
Saurischia	Theropoda			UCMP		Vertebra, ribs + frags.	Late Cret.	El Gallo	Kilmer	1957	Baja
Saurischia	Theropoda			Mex.	FMC06/053	Tooth	Late Cret.	El Gallo			Baja
Saurischia	Theropoda			UCMP		Tooth	Late Cret.	El Gallo	Colbert, E. H.	1960	Baja
Saurischia	Tyrannosauridae Fa?			LACM to Mex.	20886/WJM 7-66-2-5	Tooth	Late Cret.		Morris, W. J.	1966	Baja
Saurischia	Tyrannosauridae Fa?			LACM to Mex.	28237	Metatarsal	Late Cret.	El Gallo	Morris, W. J.	1971	Baja

GROUP	TYPE	GENUS	SPECIES	REPOSITORY	SPECIMEN NO.	ELEMENTS	AGE	FORMATION	FINDER	FIND DATE	PROVINCE
Baja turtles											
	Testudinidae			LACM	28995	Carapace frags.	Late Cret.	La Bocana Roja	Garbani, H. J.	1971	Baja
	Chelonia			LACM	101158/28992	Carapace frags.	Late Cret.	El Gallo	Garbani, H. J.	1971	Baja
	Chelonia			LACM	20878/JHG [HJG] 60	Plastron (incompl.)	Late Cret.	El Gallo	Morris, W. J.	1967	Baja
	Chelonia			LACM	20887/WJM 6-66-1-7	Carapace frags.	Late Cret.	El Gallo	Morris, W. J.	1966	Baja
		Cryptodira		LACM	42662/HJG 693	Carapace frags.	Late Cret.	El Gallo	Greenwald, M. T.	1973	Baja
		Cryptodira Sbor?		LACM	42664/HJG 693	Carapace frag.?	Late Cret.	El Gallo	Greenwald, M. T.	1973	Baja
		Cryptodira		LACM	42683/HJG 697	Vertebra cervical + limb bone	Late Cret.	El Gallo	Greenwald, M. T.	1973	Baja
		Cryptodira		LACM	42693/HJG 702	Carapace frag.	Late Cret.	El Gallo	Greenwald, M. T.	1973	Baja
		Cryptodira		LACM	57873	Carapace frag.	Late Cret.	El Gallo	Greenwald, M. T.	1973	Baja
		Cryptodira		LACM	28999	Carapace frags.	Late Cret.	El Gallo	Garbani, H. J.	1971	Baja
		Cryptodira		LACM	42673/HJG 696	Phalanx ungal	Late Cret.	El Gallo	Garbani, H. J.	1973	Baja
	Trionychidae			LACM	42663/HJG 693	Carapace frag.	Late Cret.	El Gallo	Greenwald, M. T.	1973	Baja
	Trionychidae			LACM	42692/HJG 702	Carapace frag.	Late Cret.	El Gallo	Greenwald, M. T.	1973	Baja
	Trionychidae			LACM	57874	Carapace frag.	Late Cret.	El Gallo	Greenwald, M. T.	1973	Baja
	Trionychidae			LACM	42674/HJG 696	Carapace frag.	Late Cret.	El Gallo	Garbani, H. J.	1973	Baja
	Trionychidae			LACM	42697/HJG 703	Carapace frag.	Late Cret.	El Gallo	Garbani, H. J.	1973	Baja
	Trionychidae			LACM	52455	Carapace frag.	Late Cret.	El Gallo	Morris, W. J.		Baja
Baja lizards/ snakes											
	Lacertilia			LACM	52457	Vertebra frag.	Late Cret.	El Gallo	Morris, W. J.		Baja
	Lacertilia			LACM	42708/HJG 704	Maxilla frag. w/ tooth	Late Cret.	El Gallo	Fonseca, P.	1973	Baja
	Lacertilia			LACM	42572/HJG 680	Vertebra	Late Cret.	El Gallo	Greenwald, M. T.	1973	Baja
	Lacertilia			LACM	42720/HJG 705	Phalanx, proximal end	Late Cret.	El Gallo	Greenwald, M. T.	1973	Baja
	Lacertilia			LACM	57875/HJG 705	Phalanx	Late Cret.	El Gallo	Greenwald, M. T.	1973	Baja
	Lacertilia			LACM	28994	Phalanx	Late Cret.	El Gallo	Garbani, H. J.	1971	Baja
	Lacertilia			LACM	42580/HJG 684	Phalanx	Late Cret.	El Gallo	Garbani, H. J.	1973	Baja
	Lacertilia			LACM	42581/HJG 684	Vertebra	Late Cret.	El Gallo	Garbani, H. J.	1973	Baja
	Lacertilia			LACM	42676/HJG 696	Mandible	Late Cret.	El Gallo	Garbani, H. J.	1973	Baja
	Lacertilia Sbor?			LACM	42677/HJG 696	Phalanx ungal	Late Cret.	El Gallo	Garbani, H. J.	1973	Baja
	Lacertilia			LACM	42691/HJG 701	Skeleton	Late Cret.	El Gallo	Garbani, H. J.	1973	Baja
	Lacertilia Sbor cf.			LACM	42698/HJG 703	Phalanx, proximal end	Late Cret.	El Gallo	Garbani, H. J.	1973	Baja
	Lacertilia Sbor?			LACM	42709/HJG 704	Phalanx ungal	Late Cret.	El Gallo	Garbani, H. J.	1973	Baja
	Lacertilia Sbor?			LACM	42710/HJG 704	Phalanx?	Late Cret.	El Gallo	Garbani, H. J.	1973	Baja
	Lacertilia			LACM	42718/HJG 705	Mandible, rt.	Late Cret.	El Gallo	Garbani, H. J.	1973	Baja
	Lacertilia			LACM	42721/HJG 705	Phalanx ungal frags.	Late Cret.	El Gallo	Garbani, H. J.	1973	Baja
	Squamata			LACM	58007		Late Cret.	El Gallo	Lillegraven, J. A.	1970	Baja
	Teiidae cf.			LACM	42719/HJG 705	Mandible, lft.	Late Cret.	El Gallo	Greenwald, M. T.	1973	Baja

GROUP	TYPE	GENUS	SPECIES	REPOSITORY	SPECIMEN NO.	ELEMENTS	AGE	FORMATION	FINDER	FIND DATE	PROVINCE
Lacertilia		*Polyglyphanodon*		LACM	57869	Dentary frag. w/ teeth	Late Cret.	El Gallo	Lillegraven, J. A.	1973	Baja
Lacertilia		*Polyglyphanodon*		LACM	58008	Tooth	Late Cret.	El Gallo	Lillegraven, J. A.	1973	Baja
Lacertilia		*Polyglyphanodon*		LACM	58011	Tooth + limb bone frags.	Late Cret.	El Gallo	Lillegraven, J. A.	1970	Baja
Lacertilia		*Polyglyphanodon*		LACM	57870	Limb bone frags.	Late Cret.	El Gallo	Greenwald, M. T.	1973	Baja
Lacertilia		*Polyglyphanodon*		LACM	57877	Tooth	Late Cret.	El Gallo	Greenwald, M. T.	1973	Baja
Lacertilia	Squamata			LACM	101161/28992	Mandible, lft. (lower anterior)	Late Cret.	El Gallo	Garbani, H.J.	1971	Baja
Lacertilia	Squamata			LACM	101162/28992	Vertebra	Late Cret.	El Gallo	Garbani, H.J.	1971	Baja
Lacertilia	Squamata			LACM	101166/28999	Vertebra	Late Cret.	El Gallo	Garbani, H.J.	1971	Baja

Baja Crocodilia

GROUP	TYPE	GENUS	SPECIES	REPOSITORY	SPECIMEN NO.	ELEMENTS	AGE	FORMATION	FINDER	FIND DATE	PROVINCE
Crocodilia				LACM	42577/HJG 687	Dermal scute	Late Cret.	El Gallo	Lightwardt, E.	1973	Baja
Crocodilia				LACM	42560/HJG 687	Tooth	Late Cret.	El Gallo	Greenwald, M. T.	1973	Baja
Crocodilia				LACM	42568/HJG 680	Dermal scute	Late Cret.	El Gallo	Greenwald, M. T.	1973	Baja
Crocodilia				LACM	42576/HJG 687	Tooth	Late Cret.	El Gallo	Greenwald, M. T.	1973	Baja
Crocodilia				LACM	42629/HJG 688	Tooth	Late Cret.	El Gallo	Greenwald, M. T.	1973	Baja
Crocodilia				LACM	42579/HJG 684	Tooth	Late Cret.	El Gallo	Greenwald, M. T.	1973	Baja
Crocodilia				LACM	42630/HJG 688	Dermal scute	Late Cret.	El Gallo	Greenwald, M. T.	1973	Baja
Crocodilia				LACM	42642/HJG 690	Tooth	Late Cret.	El Gallo	Greenwald, M. T.	1973	Baja
Crocodilia				LACM	42665/HJG 693	Dentary frag.	Late Cret.	El Gallo	Greenwald, M. T.	1973	Baja
Crocodilia				LACM	42666/HJG 693	Dermal scute	Late Cret.	El Gallo	Greenwald, M. T.	1973	Baja
Crocodilia				LACM	42694/HJG 702	Dermal scute	Late Cret.	El Gallo	Greenwald, M. T.	1973	Baja
Crocodilia				LACM	42695/HJG 702	Dermal scute	Late Cret.	El Gallo	Greenwald, M. T.	1973	Baja
Crocodilia				LACM	101160	Vertebra caudal	Late Cret.	El Gallo	Greenwald, M. T.	1971	Baja
Crocodilia				LACM	42562/HJG 679	Tooth	Late Cret.	El Gallo	Garbani, H.J.	1973	Baja
Crocodilia				LACM	42569/HJG 680	Tooth	Late Cret.	El Gallo	Garbani, H.J.	1973	Baja
Crocodilia				LACM	42643/HJG 690	Dermal scute	Late Cret.	El Gallo	Garbani, H.J.	1973	Baja
Crocodilia				LACM	52456	Humerus	Late Cret.	El Gallo	Morris, W. J.		Baja
Crocodilia		*Brachychampsa* Gs cf.		LACM	101159/28992	Tooth	Late Cret.	El Gallo	Garbani, H.J.	1971	Baja
Crocodilia		*Leidyosuchus*		LACM	101165/28999	Tooth	Late Cret.	El Gallo	Garbani, H.J.	1971	Baja
Crocodilia	Alligatoridae			LACM	42650/HJG 692	Skull + skeleton (anterior)	Late Cret.	El Gallo	Garbani, H.J.	1973	Baja
Crocodilia	Alligatoridae			LACM	101181/17704	Tooth	Late Cret.	El Gallo	Morris, W. J.	1966	Baja

Baja reptiles

GROUP	TYPE	GENUS	SPECIES	REPOSITORY	SPECIMEN NO.	ELEMENTS	AGE	FORMATION	FINDER	FIND DATE	PROVINCE
Reptilia				LACM	42566/HJG 679	Innominate frags.	Late Cret.	El Gallo	Fonseca, P.	1973	Baja
Reptilia				LACM	42573/HJG 680	Limb bone	Late Cret.	El Gallo	Greenwald, M. T.	1973	Baja
Reptilia				LACM	42582/HJG 684	Vertebra	Late Cret.	El Gallo	Greenwald, M. T.	1973	Baja
Reptilia				LACM	42668/HJG 693	Vertebra	Late Cret.	El Gallo	Greenwald, M. T.	1973	Baja
Reptilia				LACM	42696/HJG 702	Limb bone	Late Cret.	El Gallo	Greenwald, M. T.	1973	Baja

GROUP	TYPE	GENUS	SPECIES	REPOSITORY	SPECIMEN NO.	ELEMENTS	AGE	FORMATION	FINDER	FIND DATE	PROVINCE
Reptilia				LACM	101178	Coprolites	Late Cret.	El Gallo	Greenwald, M. T.	1971	Baja
Reptilia				LACM	29001	Skull frag.	Late Cret.	El Gallo	Garbani, H.J.	1971	Baja
Reptilia				LACM	42575/HJG 682	Skull frag.	Late Cret.	El Gallo	Garbani, H.J.	1973	Baja
Reptilia				LACM	42589/HJG 687	Skull (posterior), frags.	Late Cret.	El Gallo	Garbani, H.J.	1973	Baja
Reptilia				LACM	42645/HJG 690	Tooth	Late Cret.	El Gallo	Garbani, H.J.	1973	Baja
Reptilia				LACM	42679	Tooth?	Late Cret.	El Gallo	Garbani, H.J.	1973	Baja
Reptilia Cf?				LACM	42699/HJG 703	Limb bone	Late Cret.	El Gallo	Garbani, H.J.	1973	Baja
Reptilia Cf?				LACM	42701/HJG 703	Dermal bone	Late Cret.	El Gallo	Garbani, H.J.	1973	Baja
Reptilia Cf?				LACM	42711/HJG 704	Limb bone, distal end	Late Cret.	El Gallo	Garbani, H.J.	1973	Baja
Reptilia Cf?				LACM	42722/HJG 705	Innominate acetabulum?	Late Cret.	El Gallo	Garbani, H.J.	1973	Baja
Reptilia				LACM	101175/28892	Limb bones	Late Cret.	El Gallo	Garbani, H.J.	1971	Baja
Reptilia				LACM	101176/28992	Phalanx ungal	Late Cret.	El Gallo	Garbani, H.J.	1971	Baja
Reptilia				LACM	101177/28993	Eggshell frags.	Late Cret.	El Gallo	Garbani, H.J.	1971	Baja
Reptilia				LACM	101179/28997		Late Cret.	El Gallo	Garbani, H.J.	1971	Baja
Reptilia				LACM	101180/28999	Mandible (lower posterior)	Late Cret.	El Gallo	Garbani, H.J.	1971	Baja
Baja birds											
Aves		*Alexornis*		LACM	33213	Rt. humerus distal, lft. scapula, lft. coracoid, rt. ulna, rt. tibiotarsus, lft. femur	Late Cret.	La Bocana Roja	Garbani, H.J. & Loewe		Baja
FOSSIL REPTILES FROM ALTA CALIFORNIA											
California dinosaurs											
dinosaur				SBCM		Tracks	Jurassic	Aztec	Evans, J. R.	1958	San Bernardino
dinosaur				SDNHM	67367	Metapodial frag.	Late Cret.	Pt. Loma	Riney, B.	1987	San Diego
dinosaur?				SC	VR81	Metatarsal?, distal end	Jurassic	Great Valley Group	Jensen, J., Jr.	1998	Tehama
dinosaur?				SC	82	Rib, proximal end	Late Jur.	Trail	Christe, G.	1997	Plumas
dinosaur?				SC	83	Neural spine from vertebra?	Late Jur.	Trail	Christe, G.	1995	Plumas
dinosaur?				SC	84	Fragments	Late Jur.	Trail	Christe, G.	1995	Plumas
Ornithischia	Ankylosauridae			SDNHM	33909	Skeleton, postcranial (partial)	Late Cret.	Pt. Loma	Riney, B.	1987	San Diego
Ornithischia	Hypsilophodont			SC	VR04	Foot bones (lft. hind leg)	Early Cret.	Budden Canyon	Hilton, R.	1991	Shasta
Ornithischia	Nodosaur			SDNHM	33909	Skeleton	Campanian/Maastrichtian	Pt. Loma	Riney, B.	1987	Orange
Ornithischia	Hadrosauridae	*Saurolophus*		LACM/CIT	2760	Skull, mandibles + limb bones	Late Cret.	Moreno	Smith, B. & Leard, R. M. et al.	1939	Fresno
Ornithischia	Hadrosauridae	*Saurolophus*		LACM/CIT	2852	Skull + skeleton	Late Cret.	Moreno	Drescher, A. B. & Leard, R. M.	1943	Fresno
Ornithischia	Hadrosauridae			LACM/CIT	5219	Maxilla frag. w/ teeth	Late Cret.	Ladd/Williams	Moore, B.		Orange

GROUP	TYPE	GENUS	SPECIES	REPOSITORY	SPECIMEN NO.	ELEMENTS	AGE	FORMATION	FINDER	FIND DATE	PROVINCE
Ornithischia	Hadrosauridae			OCNHF		Tibia, distal end	Late Cret.	Ladd	Hansen, R. D.	1992	Orange
Ornithischia	Hadrosauridae			OCNHF	1785	Frontal cranium	Late Cret.	Ladd	Drachuk, R.	1978	Orange
Ornithischia	Hadrosauridae			SDNHM	67368	Dentary, rt., w/ teeth	Late Cret.	Pt. Loma	Case, L.	1987	San Diego
Ornithischia	Hadrosauridae			SDNHM	25342	Femur	Late Cret.	Pt. Loma	Riney, B.	1983	San Diego
Ornithischia	Hadrosauridae			SDNHM		Vertebrae cervical	Late Cret.	Pt. Loma	Riney, B. (as child)	1967	San Diego
Ornithischia	Hadrosauridae			SDNHM	66640	Vertebrae (13)	Late Cret.	Pt. Loma	Riney, B.	1986	San Diego
Ornithischia	Hadrosauridae			UCMP	32944	Several vertebrae	Late Cret.	Moreno	Bennison, A. & UC party	1936	Stanislaus
Ornithischia	Hadrosauridae			UCMP	137237	Partial skeleton	Late Cret.	Moreno	Ervin, S.; Staebler, A. & Staebler, C. N.	1985	Fresno
Ornithischia	Hadrosauridae			UCMP	137242	Skull, mandible, vertebrae (3) + rib frags.	Late Cret.	Moreno	Staebler, C. N.	1982	Fresno
Ornithischia	Hadrosauridae			YPM-PU	19333	Maxilla frag., lft.	E. or Lt. Cret.	Great Valley Group	Case, G.	1966	Tehama
Ornithischia	Hadrosauridae			ETC = OCPC	GC958119 = 21913, 22265–22271	2 cervical vertebrae; rib, 2 phalanges	Late Cret.	Williams/Ladd	Calvano, G.	1996	Orange
Ornithischia	Hadrosauridae?			ETC = OCPC	GC9781.130 = 22109	Frags.	Late Cret.	Williams	Calvano, G.	1996	Orange
Ornithischia	Hadrosauridae?			OCPC	22144–22146, 22495–22496	Limb bone frags., phalanges	Late Cret.	Williams	Peck, P.	1996	Orange
Saurischia	coelurosaur	cf. *Grallator*		SBCM		Tracks	Jurassic	Aztec	Evans, J. R.	1958	San Bernardino
Saurischia	coelurosaur	cf. *Anchisauripus*		SBCM		Tracks	Jurassic	Aztec	Evans, J. R.	1958	San Bernardino
Saurischia	coelurosaur			SBCM		Tracks	Jurassic	Aztec	Evans, J. R.	1958	San Bernardino
Saurischia	Theropod			SC	VRD57	Limb bone frag.	Late Cret.	Chico	Antuzzi, P.	1995	Placer
CALIFORNIA TRIASSIC MARINE REPTILES											
Thalattosauria											
Thalattosauria	Thalattosauridae	*Nectosaurus*		UCMP	10760	Pterygoid w/ teeth	Late Triassic	Hosselkus	Merriam, J. C.	1906	Shasta
Thalattosauria	Thalattosauridae	*Nectosaurus*		UCMP	10766		Late Triassic	Hosselkus	UC party	1906	Shasta
Thalattosauria	Thalattosauridae	*Nectosaurus*		UCMP	10767		Late Triassic	Hosselkus	UC party	1906	Shasta
Thalattosauria	Thalattosauridae	*Nectosaurus*		UCMP	10769		Late Triassic	Hosselkus	UC party	1906	Shasta
Thalattosauria	Thalattosauridae	*Nectosaurus*		UCMP	10770	Pterygoid	Late Triassic	Hosselkus	UC party	1906	Shasta
Thalattosauria	Thalattosauridae	*Nectosaurus*		UCMP	10771		Late Triassic	Hosselkus	Merriam, J. C.	1906	Shasta
Thalattosauria	Thalattosauridae	*Nectosaurus*		UCMP	10772	Limb bone	Late Triassic	Hosselkus	Merriam, J. C.	1906	Shasta
Thalattosauria	Thalattosauridae	*Nectosaurus*		UCMP	10775	Limb bone	Late Triassic	Hosselkus	Merriam, J. C.	1906	Shasta
Thalattosauria	Thalattosauridae	*Nectosaurus*		UCMP	10778	Premaxilla or prevomer w/ teeth	Late Triassic	Hosselkus	Merriam, J. C.	1906	Shasta
Thalattosauria	Thalattosauridae	*Nectosaurus*		UCMP	10780		Late Triassic	Hosselkus	Merriam, J. C.	1906	Shasta
Thalattosauria	Thalattosauridae	*Nectosaurus*		UCMP	10781		Late Triassic	Hosselkus	Merriam, J. C.	1906	Shasta
Thalattosauria	Thalattosauridae	*Nectosaurus*		UCMP	10782		Late Triassic	Hosselkus	Merriam, J. C.	1906	Shasta
Thalattosauria	Thalattosauridae	*Nectosaurus*		UCMP	10784		Late Triassic	Hosselkus	Merriam, J. C.	1906	Shasta

GROUP	TYPE	GENUS	SPECIES	REPOSITORY	SPECIMEN NO.	ELEMENTS	AGE	FORMATION	FINDER	FIND DATE	PROVINCE
Thalattosauria	Thalattosauridae	Nectosaurus		UCMP	10787		Late Triassic	Hosselkus	Merriam, J. C.	1906	Shasta
Thalattosauria	Thalattosauridae	Nectosaurus		UCMP	10788	Vertebra	Late Triassic	Hosselkus	Merriam, J. C.	1906	Shasta
Thalattosauria	Thalattosauridae	Nectosaurus		UCMP	10791		Late Triassic	Hosselkus	Merriam, J. C.	1906	Shasta
Thalattosauria	Thalattosauridae	Nectosaurus		UCMP	10792	Bone frag.	Late Triassic	Hosselkus	Merriam, J. C.	1906	Shasta
Thalattosauria	Thalattosauridae	Nectosaurus		UCMP	10793		Late Triassic	Hosselkus	Merriam, J. C.	1906	Shasta
Thalattosauria	Thalattosauridae	Nectosaurus		UCMP	10794		Late Triassic	Hosselkus	Merriam, J. C.	1906	Shasta
Thalattosauria	Thalattosauridae	Nectosaurus		UCMP	10797	Mandible + epipodial	Late Triassic	Hosselkus	Merriam, J. C.	1906	Shasta
Thalattosauria	Thalattosauridae	Nectosaurus		UCMP	10800	Propodial	Late Triassic	Hosselkus	Merriam, J. C.	1908	Shasta
Thalattosauria	Thalattosauridae	Nectosaurus		UCMP	10799	Rib?	Late Triassic	Hosselkus	Merriam, J. C.	1906	Shasta
Thalattosauria	Thalattosauridae	Nectosaurus		UCMP	10809	Bone frags.	Late Triassic	Hosselkus	Merriam, J. C.	1906	Shasta
Thalattosauria	Thalattosauridae	Nectosaurus		UCMP	10814	Bone frag.	Late Triassic	Hosselkus	Merriam, J. C.	1906	Shasta
Thalattosauria	Thalattosauridae	Nectosaurus		UCMP	10816	Bone frags.	Late Triassic	Hosselkus	Merriam, J. C.	1906	Shasta
Thalattosauria	Thalattosauridae	Nectosaurus		UCMP	10817	Pterygoid w/ teeth	Late Triassic	Hosselkus	Merriam, J. C.	1906	Shasta
Thalattosauria	Thalattosauridae	Nectosaurus		UCMP	10620	Dentary	Late Triassic	Hosselkus	Merriam, J. C.	1905	Shasta
Thalattosauria	Thalattosauridae	Nectosaurus		UCMP	10818	Bone frag.	Late Triassic	Hosselkus	Merriam, J. C.	1906	Shasta
Thalattosauria	Thalattosauridae	Nectosaurus		UCMP	10704	Teeth	Late Triassic	Hosselkus	UC party	1910	Shasta
Thalattosauria	Thalattosauridae	Nectosaurus		UCMP	10764	Mandible w/ teeth	Late Triassic	Hosselkus	UC party	1906	Shasta
Thalattosauria	Thalattosauridae	Nectosaurus		UCMP	10765	Teeth	Late Triassic	Hosselkus	UC party	1906	Shasta
Thalattosauria	Thalattosauridae	Nectosaurus		UCMP	10768	Ischium + pubis	Late Triassic	Hosselkus	UC party	1906	Shasta
Thalattosauria	Thalattosauridae	Nectosaurus		UCMP	10776	Vertebra	Late Triassic	Hosselkus	Merriam, J. C.	1906	Shasta
Thalattosauria	Thalattosauridae	Nectosaurus	halius	UCMP	10757	Rib + vertebra frag.	Late Triassic	Hosselkus	Merriam, J. C.	1906	Shasta
Thalattosauria	Thalattosauridae	Nectosaurus	halius?	UCMP	10758	Frontal tooth	Late Triassic	Hosselkus	Merriam, J. C.	1906	Shasta
Thalattosauria	Thalattosauridae	Nectosaurus	halius?	UCMP	10759	Frontal bone	Late Triassic	Hosselkus	Merriam, J. C.	1906	Shasta
Thalattosauria	Thalattosauridae	Nectosaurus	halius	UCMP	10773	Bone frags.	Late Triassic	Hosselkus	Merriam, J. C.	1906	Shasta
Thalattosauria	Thalattosauridae	Nectosaurus	halius	UCMP	9124	Frontoparietal frag. + lower mandible	Late Triassic	Hosselkus			Shasta
Thalattosauria	Thalattosauridae	Nectosaurus	halius	UCMP	9125	Post mandible	Late Triassic	Hosselkus			Shasta
Thalattosauria	Thalattosauridae	Nectosaurus	halius?	UCMP	10626	Pterygoid w/ teeth + propodial	Late Triassic	Hosselkus	Merriam, J. C.	1906	Shasta
Thalattosauria	Thalattosauridae	Nectosaurus	halius	UCMP	10627	Vertebra frag.	Late Triassic	Hosselkus	Merriam, J. C.	1906	Shasta
Thalattosauria	Thalattosauridae	Nectosaurus	halius	UCMP	10753	Maxilla w/ teeth	Late Triassic	Hosselkus	Merriam, J. C.	1906	Shasta
Thalattosauria	Thalattosauridae	Nectosaurus	halius?	UCMP	10755	Podial	Late Triassic	Hosselkus	Merriam, J. C.	1906	Shasta
Thalattosauria	Thalattosauridae	Nectosaurus	halius	UCMP	10756	Bone frag.	Late Triassic	Hosselkus	Merriam, J. C.	1906	Shasta
Thalattosauria	Thalattosauridae	Nectosaurus	halius	UCMP	10779	Vertebra frag.	Late Triassic	Hosselkus	Merriam, J. C.	1906	Shasta
Thalattosauria	Thalattosauridae	Nectosaurus	halius?	UCMP	10783	Epipodial	Late Triassic	Hosselkus	Merriam, J. C.	1906	Shasta
Thalattosauria	Thalattosauridae	Nectosaurus	halius	UCMP	10801	Scapula	Late Triassic	Hosselkus	Merriam, J. C.	1906	Shasta
Thalattosauria	Thalattosauridae	Nectosaurus	halius?	UCMP	10803	Scapula	Late Triassic	Hosselkus	Merriam, J. C.	1906	Shasta
Thalattosauria	Thalattosauridae	Nectosaurus	halius	UCMP	10621	Pterygoid or palatine	Late Triassic	Hosselkus	UC party		Shasta

GROUP	TYPE	GENUS	SPECIES	REPOSITORY	SPECIMEN NO.	ELEMENTS	AGE	FORMATION	FINDER	FIND DATE	PROVINCE
Thalattosauria	Thalattosauridae	*Nectosaurus*	*halius*	UCMP	10625		Late Triassic	Hosselkus	UC party		Shasta
Thalattosauria	Thalattosauridae	*Nectosaurus*	*halius*	UCMP	136581	Mandible w/ teeth	Late Triassic	Hosselkus	Merriam, J. C.	1905	Shasta
Thalattosauria	Thalattosauridae	*Nectosaurus*	*halius*	UCMP	113686	Maxilla	Late Triassic	Hosselkus	Alexander, A. M.	1906	Shasta
Thalattosauria	Thalattosauridae	*Nectosaurus*	*halius*	UCMP	10774	Vertebra dorsal, vertebra caudal + frags.	Late Triassic	Hosselkus	Merriam, J. C.	1906	Shasta
Thalattosauria	Thalattosauridae	*Thalattosaurus*	*alexandrae*	UCMP	10815	Teeth	Late Triassic	Hosselkus	Merriam, J. C.	1906	Shasta
Thalattosauria	Thalattosauridae	*Thalattosaurus*	*alexandrae*	UCMP	9045	Pubis	Late Triassic	Hosselkus			Shasta
Thalattosauria	Thalattosauridae	*Thalattosaurus*	*alexandrae*	UCMP	9044	Pubis?	Late Triassic	Hosselkus			Shasta
Thalattosauria	Thalattosauridae	*Thalattosaurus*	*alexandrae*	UCMP	9084	Skull frag., vertebra, rib + limb bones	Late Triassic	Hosselkus			Shasta
Thalattosauria	Thalattosauridae	*Thalattosaurus*	*alexandrae*	UCMP	9122	Skull frag. w/ teeth	Late Triassic	Hosselkus	Merriam, J. C.		Shasta
Thalattosauria	Thalattosauridae	*Thalattosaurus*	*alexandrae*	UCMP	8123	Premaxilla	Late Triassic	Hosselkus			Shasta
Thalattosauria	Thalattosauridae	*Thalattosaurus*	*alexandrae*	UCMP	9085	Skull frags., limb bones + vertebrae	Late Triassic	Hosselkus			Shasta
Thalattosauria	Thalattosauridae	*Thalattosaurus*	*alexandrae*	UCMP	9121	Mandible w/ teeth	Late Triassic	Hosselkus			Shasta
Thalattosauria	Thalattosauridae	*Thalattosaurus*	*alexandrae*	UCMP	9126	Tooth	Late Triassic	Hosselkus			Shasta
Thalattosauria	Thalattosauridae	*Thalattosaurus*	*alexandrae*	UCMP	39226	Mandible, rt. upper + lower mandible	Late Triassic	Hosselkus			Shasta
Thalattosauria	Thalattosauridae	*Thalattosaurus*	*alexandrae*	UCMP	10628	Teeth	Late Triassic	Hosselkus	UC party		Shasta
Thalattosauria	Thalattosauridae	*Thalattosaurus*	*alexandrae*	UCMP	9123	Premaxilla (posterior)	Late Triassic	Hosselkus			Shasta
Thalattosauria	Thalattosauridae	*Thalattosaurus*	*alexandrae*	UCMP	11228	Mandible + teeth	Late Triassic	Hosselkus	UC party		Shasta
Thalattosauria	Thalattosauridae	*Thalattosaurus*	*alexandrae*	UCMP	39266	Skull	Late Triassic	Hosselkus	Merriam, J. C.		Shasta
Thalattosauria	Thalattosauridae	*Thalattosaurus*	*shastensis*	USNM	10926	Skull	Late Triassic	Hosselkus	Stanton, T. W.	1902	Shasta
Thalattosauria	Thalattosauridae	*Thalattosaurus*	*shastensis*	UCMP	9120	Skull, lft. and rt. mandible, teeth, vertebra + limb bones (anterior)	Late Triassic	Hosselkus			Shasta
Thalattosauria	Thalattosauridae	*Thalattosaurus*	*shastensis*	UCMP	9220	Mandible w/ teeth	Late Triassic	Hosselkus	Smith, J. P.		Shasta
Thalattosauria	Thalattosauridae	*Thalattosaurus*	*perrini*	CAS			Late Triassic	Hosselkus			
Ichthyosauria	Shastasauridae	*Merriamia*	*zitteli*	UCMP	8099	Skull frags., limb bones + pectra	Late Triassic	Hosselkus	Alexander, A. M.		Shasta
Ichthyosauria	Shastasauridae	*Shastasaurus*		UCMP	9019		Late Triassic	Hosselkus			Shasta
Ichthyosauria	Shastasauridae	*Shastasaurus*		UCMP	9022		Late Triassic	Hosselkus	Alexander, A. M.	1903	Shasta
Ichthyosauria	Shastasauridae	*Shastasaurus*		UCMP	9023	Vertebra	Late Triassic	Hosselkus	Alexander, A. M.	1903	Shasta
Ichthyosauria	Shastasauridae	*Shastasaurus*		UCMP	9029	Bone frags.	Late Triassic	Hosselkus	Alexander, A. M. & Furlong, E.L.	1903	Shasta
Ichthyosauria	Shastasauridae	*Shastasaurus*		UCMP	9032		Late Triassic	Hosselkus	Alexander, A. M. & Furlong, E.L.	1903	Shasta
Ichthyosauria	Shastasauridae	*Shastasaurus*		UCMP	9034		Late Triassic	Hosselkus	Alexander, A. M. & Furlong, E.L.	1903	Shasta

GROUP	TYPE	GENUS	SPECIES	REPOSITORY	SPECIMEN NO.	ELEMENTS	AGE	FORMATION	FINDER	FIND DATE	PROVINCE
Ichthyosauria	Shastasauridae	*Shastasaurus*		UCMP	9041		Late Triassic	Hosselkus	Alexander, A. M. & Furlong, E. L.	1903	Shasta
Ichthyosauria	Shastasauridae	*Shastasaurus*		UCMP	9050	Vertebrae + ribs	Late Triassic	Hosselkus	Alexander, A. M. & Furlong, E. L.	1903	Shasta
Ichthyosauria	Shastasauridae	*Shastasaurus*		UCMP	9071	Skull frags. + vertebra	Late Triassic	Hosselkus	Alexander, A. M. & Furlong, E. L.	1903	Shasta
Ichthyosauria	Shastasauridae	*Shastasaurus*		UCMP	9072	Vertebra	Late Triassic	Hosselkus	Alexander, A. M. & Furlong, E. L.	1903	Shasta
Ichthyosauria	Shastasauridae	*Shastasaurus*		UCMP	9073	Vertebra	Late Triassic	Hosselkus	Alexander, A. M. & Furlong, E. L.	1903	Shasta
Ichthyosauria	Shastasauridae	*Shastasaurus*		UCMP	9089	Ribs	Late Triassic	Hosselkus	Alexander, A. M. & Furlong, E. L.	1903	Shasta
Ichthyosauria	Shastasauridae	*Shastasaurus*		UCMP	9090		Late Triassic	Hosselkus	Alexander, A. M. & Furlong, E. L.	1903	Shasta
Ichthyosauria	Shastasauridae	*Shastasaurus*		UCMP	9095	Vomer? w/ teeth	Late Triassic	Hosselkus	Alexander, A. M. & Furlong, E. L.	1903	Shasta
Ichthyosauria	Shastasauridae	*Shastasaurus*		UCMP	9096	Vomer?	Late Triassic	Hosselkus	Alexander, A. M. & Furlong, E. L.	1903	Shasta
Ichthyosauria	Shastasauridae	*Shastasaurus*		UCMP	9097	Vomer?	Late Triassic	Hosselkus	Alexander, A. M. & Furlong, E. L.	1903	Shasta
Ichthyosauria	Shastasauridae	*Shastasaurus*		UCMP	9099	Vomer? w/ teeth	Late Triassic	Hosselkus	Alexander, A. M. & Furlong, E. L.	1903	Shasta
Ichthyosauria	Shastasauridae	*Shastasaurus*		UCMP	9101	Skull frag.	Late Triassic	Hosselkus	Alexander, A. M. & Furlong, E. L.	1903	Shasta
Ichthyosauria	Shastasauridae	*Shastasaurus*		UCMP	9109	Vomer	Late Triassic	Hosselkus	Alexander, A. M. & Furlong, E. L.	1903	Shasta
Ichthyosauria	Shastasauridae	*Shastasaurus*		UCMP	9110	Tooth	Late Triassic	Hosselkus	Alexander, A. M. & Furlong, E. L.	1903	Shasta
Ichthyosauria	Shastasauridae	*Shastasaurus*		UCMP	9020	Skeleton	Late Triassic	Hosselkus	Alexander, A. M.	1903	Shasta
Ichthyosauria	Shastasauridae	*Shastasaurus*		UCMP	9021	Skeleton	Late Triassic	Hosselkus	Alexander, A. M.	1903	Shasta
Ichthyosauria	Shastasauridae	*Shastasaurus*		UCMP	9025		Late Triassic	Hosselkus	Alexander, A. M. & Furlong, E. L.	1903	Shasta
Ichthyosauria	Shastasauridae	*Shastasaurus*		UCMP	9026		Late Triassic	Hosselkus	Alexander, A. M. & Furlong, E. L.	1903	Shasta
Ichthyosauria	Shastasauridae	*Shastasaurus*		UCMP	9027	Vertebra	Late Triassic	Hosselkus	Alexander, A. M. & Furlong, E. L.	1903	Shasta
Ichthyosauria	Shastasauridae	*Shastasaurus*		UCMP	9028	Vertebra	Late Triassic	Hosselkus	Alexander, A. M. & Furlong, E. L.	1903	Shasta
Ichthyosauria	Shastasauridae	*Shastasaurus*		UCMP	9031	Vertebrae + ribs	Late Triassic	Hosselkus	Alexander, A. M. & Furlong, E. L.	1903	Shasta
Ichthyosauria	Shastasauridae	*Shastasaurus*		UCMP	9038		Late Triassic	Hosselkus	Alexander, A. M. & Furlong, E. L.	1903	Shasta
Ichthyosauria	Shastasauridae	*Shastasaurus*		UCMP	9048		Late Triassic	Hosselkus	Alexander, A. M. & Furlong, E. L.	1903	Shasta

GROUP	TYPE	GENUS	SPECIES	REPOSITORY	SPECIMEN NO.	ELEMENTS	AGE	FORMATION	FINDER	FIND DATE	PROVINCE
Ichthyosauria	Shastasauridae	*Shastasaurus*		UCMP	9049	Vertebrae + ribs	Late Triassic	Hosselkus	Alexander, A. M. & Furlong, E. L.	1903	Shasta
Ichthyosauria	Shastasauridae	*Shastasaurus*		UCMP	9609	Skeleton (incompl.)	Late Triassic	Hosselkus			Shasta
Ichthyosauria	Shastasauridae	*Shastasaurus*		UCMP	10638	Vertebra	Late Triassic	Hosselkus	UC party	1903	Shasta
Ichthyosauria	Shastasauridae	*Shastasaurus*		UCMP	9030		Late Triassic	Hosselkus	Alexander, A. M. & Furlong, E. L.	1903	Shasta
Ichthyosauria	Shastasauridae	*Shastasaurus*		UCMP	9035	Vertebra	Late Triassic	Hosselkus	Alexander, A. M. & Furlong, E. L.	1903	Shasta
Ichthyosauria	Shastasauridae	*Shastasaurus*		UCMP	9046		Late Triassic	Hosselkus	Alexander, A. M. & Furlong, E. L.	1903	Shasta
Ichthyosauria	Shastasauridae	*Shastasaurus*		UCMP	9047		Late Triassic	Hosselkus	Alexander, A. M. & Furlong, E. L.	1903	Shasta
Ichthyosauria	Shastasauridae	*Shastasaurus*		UCMP	9091		Late Triassic	Hosselkus	Alexander, A. M. & Furlong, E. L.	1903	Shasta
Ichthyosauria	Shastasauridae	*Shastasaurus*		UCMP	9051		Late Triassic	Hosselkus	Alexander, A. M.	1903	Shasta
Ichthyosauria	Shastasauridae	*Shastasaurus*		UCMP	9052	Vertebra	Late Triassic	Hosselkus	Alexander, A. M.	1903	Shasta
Ichthyosauria	Shastasauridae	*Shastasaurus*		UCMP	9053	Vertebra	Late Triassic	Hosselkus	Alexander, A. M.	1903	Shasta
Ichthyosauria	Shastasauridae	*Shastasaurus*		UCMP	9054	Vertebra	Late Triassic	Hosselkus	Alexander, A. M.	1903	Shasta
Ichthyosauria	Shastasauridae	*Shastasaurus*		UCMP	9055	Mandible, lower (frag.)	Late Triassic	Hosselkus	Alexander, A. M.	1903	Shasta
Ichthyosauria	Shastasauridae	*Shastasaurus*		UCMP	9056	Vertebra	Late Triassic	Hosselkus	Alexander, A. M.	1903	Shasta
Ichthyosauria	Shastasauridae	*Shastasaurus*		UCMP	9057	Limb bone	Late Triassic	Hosselkus	Alexander, A. M.	1903	Shasta
Ichthyosauria	Shastasauridae	*Shastasaurus*		UCMP	9058	Limb bone	Late Triassic	Hosselkus	Alexander, A. M.	1903	Shasta
Ichthyosauria	Shastasauridae	*Shastasaurus*		UCMP	9059		Late Triassic	Hosselkus	Alexander, A. M.	1903	Shasta
Ichthyosauria	Shastasauridae	*Shastasaurus*		UCMP	9060	Vertebra	Late Triassic	Hosselkus	Alexander, A. M.	1903	Shasta
Ichthyosauria	Shastasauridae	*Shastasaurus*		UCMP	9062	Vertebra	Late Triassic	Hosselkus	Alexander, A. M.	1903	Shasta
Ichthyosauria	Shastasauridae	*Shastasaurus*		UCMP	9063	Vertebra	Late Triassic	Hosselkus	Alexander, A. M.	1903	Shasta
Ichthyosauria	Shastasauridae	*Shastasaurus*		UCMP	9064	Vertebra	Late Triassic	Hosselkus	Alexander, A. M.	1903	Shasta
Ichthyosauria	Shastasauridae	*Shastasaurus*		UCMP	9065	Mandible, lower, w/ teeth + vertebra	Late Triassic	Hosselkus	Alexander, A. M.	1903	Shasta
Ichthyosauria	Shastasauridae	*Shastasaurus*		UCMP	9066	Vertebra	Late Triassic	Hosselkus	Alexander, A. M.	1903	Shasta
Ichthyosauria	Shastasauridae	*Shastasaurus*		UCMP	9067	Limb bone	Late Triassic	Hosselkus	Alexander, A. M.	1903	Shasta
Ichthyosauria	Shastasauridae	*Shastasaurus*		UCMP	9068	Vertebra	Late Triassic	Hosselkus	Alexander, A. M.	1903	Shasta
Ichthyosauria	Shastasauridae	*Shastasaurus*		UCMP	9069	Vertebra	Late Triassic	Hosselkus	Alexander, A. M.	1903	Shasta
Ichthyosauria	Shastasauridae	*Shastasaurus*		UCMP	9070		Late Triassic	Hosselkus	Alexander, A. M.	1903	Shasta
Ichthyosauria	Shastasauridae	*Shastasaurus*		UCMP	9079		Late Triassic	Hosselkus	Alexander, A. M.	1903	Shasta
Ichthyosauria	Shastasauridae	*Shastasaurus*		UCMP	9092	Mandible, lower	Late Triassic	Hosselkus	Alexander, A. M. & Furlong, E. L.	1903	Shasta
Ichthyosauria	Shastasauridae	*Shastasaurus*		UCMP	9093		Late Triassic	Hosselkus	Alexander, A. M. & Furlong, E. L.	1903	Shasta

GROUP	TYPE	GENUS	SPECIES	REPOSITORY	SPECIMEN NO.	ELEMENTS	AGE	FORMATION	FINDER	FIND DATE	PROVINCE
Ichthyosauria	Shastasauridae	Shastasaurus		UCMP	9102	Mandible, lower, w/ tooth	Late Triassic	Hosselkus	Alexander, A. M. & Furlong, E. L.	1903	Shasta
Ichthyosauria	Shastasauridae	Shastasaurus		UCMP	9103	Tooth	Late Triassic	Hosselkus	Alexander, A. M. & Furlong, E. L.	1903	Shasta
Ichthyosauria	Shastasauridae	Shastasaurus		UCMP	9104		Late Triassic	Hosselkus	Alexander, A. M. & Furlong, E. L.	1903	Shasta
Ichthyosauria	Shastasauridae	Shastasaurus		UCMP	9105	Vertebra + bone frags.	Late Triassic	Hosselkus	Alexander, A. M. & Furlong, E. L.	1903	Shasta
Ichthyosauria	Shastasauridae	Shastasaurus		UCMP	9108	Tooth	Late Triassic	Hosselkus	Alexander, A. M. & Furlong, E. L.	1903	Shasta
Ichthyosauria	Shastasauridae	Shastasaurus		UCMP	31537		Late Triassic	Hosselkus	Alexander, A. M.	1903	Shasta
Ichthyosauria	Shastasauridae	Shastasaurus		UCMP	9061	Limb bone	Late Triassic	Hosselkus	Alexander, A. M.	1903	Shasta
Ichthyosauria	Shastasauridae	Shastasaurus		UCMP	9115	Vertebra	Late Triassic	Hosselkus			Shasta
Ichthyosauria	Shastasauridae	Shastasaurus		UCMP	9024	Vertebra	Late Triassic	Hosselkus	Alexander, A. M.	1903	Shasta
Ichthyosauria	Shastasauridae	Shastasaurus	pacificus	CAS		Scapula	Late Triassic	Hosselkus	Smith, J. P.		Shasta
Ichthyosauria	Shastasauridae	Shastasaurus	pacificus	CAS		Scapula frag.	Late Triassic	Hosselkus	Smith, J. P.		Shasta
Ichthyosauria	Shastasauridae	Shastasaurus	pacificus	UCMP	9077	Vertebrae + ribs	Late Triassic	Hosselkus	Alexander, A. M. & Furlong, E. L.	1903	Shasta
Ichthyosauria	Shastasauridae	Shastasaurus	careyi	UCMP	9614	Humerus + radius	Late Triassic	Hosselkus	Alexander, A. M.	1903	Shasta
Ichthyosauria	Shastasauridae	Shastasaurus	careyi	UCMP	9075	Vertebrae + rib, proximal end	Late Triassic	Hosselkus	Alexander, A. M. & Furlong, E. L.	1903	Shasta
Ichthyosauria	Shastasauridae	Shastasaurus (new, Shonosaurus sp. Motani 1999)	careyi	UCMP	9080	Vertebra	Late Triassic	Hosselkus	Alexander, A. M. & Furlong, E. L.	1903	Shasta
Ichthyosauria	Shastasauridae	Shastasaurus	osmonti	UCMP	9076	Vertebrae, ribs + limb bones	Late Triassic	Hosselkus	Alexander, A. M. & Furlong, E. L.	1903	Shasta
Ichthyosauria	Shastasauridae	Shastasaurus	osmonti	UCMP	9608	Skeleton, dorsal (posterior)	Late Triassic	Hosselkus			Shasta
Ichthyosauria	Shastasauridae	Shastasaurus	alexandrae	UCMP	9017	Skeleton (anterior)	Late Triassic	Hosselkus			Shasta
Ichthyosauria	Shastasauridae	Shastasaurus	altispinus	UCMP	9083	Vertebrae, ribs + limb bone frags.	Late Triassic	Hosselkus			Shasta
Ichthyosauria	Shastasauridae	Toretocnemus	californicus	UCMP	10998	Skeleton	Late Triassic	Hosselkus	Merriam, J. C.		Shasta
Ichthyosauria	Shastasauridae	Toretocnemus		UCMP	8100	Skeleton (incompl.)	Late Triassic	Hosselkus			Shasta
Ichthyosauria	Shastasauridae	Toretocnemus	perrini	UCMP	9082	Pectoral girdle + forelimbs	Late Triassic	Hosselkus			Shasta
Ichthyosauria	Shastasauridae	Toretocnemus	perrini	UCMP	9086	Vertebra + limb bones	Late Triassic	Hosselkus			Shasta
Ichthyosauria	Shastasauridae	Toretocnemus	perrini	UCMP	9112	Vertebra	Late Triassic	Hosselkus			Shasta
Ichthyosauria	Shastasauridae	Toretocnemus	perrini	UCMP	9119	Skeleton	Late Triassic	Hosselkus			Shasta
Ichthyosauria	Shastasauridae	Toretocnemus	perrini	UCMP	9111	Vertebra caudal	Late Triassic	Hosselkus	Sherman, O. B.	1903	Shasta
Ichthyosauria	Shastasauridae	Toretocnemus	perrini	UCMP	9113	Vertebra	Late Triassic	Hosselkus			Shasta
Ichthyosauria	Shastasauridae	Toretocnemus	perrini	UCMP	9114	Rib	Late Triassic	Hosselkus			Shasta
Ichthyosauria	Shastasauridae	Toretocnemus	perrini	UCMP	9116	Vertebra + rib	Late Triassic	Hosselkus			Shasta

GROUP	TYPE	GENUS	SPECIES	REPOSITORY	SPECIMEN NO.	ELEMENTS	AGE	FORMATION	FINDER	FIND DATE	PROVINCE
Ichthyosauria	Shastasauridae	Toretocnemus	perrini	UCMP	9117	Vertebra	Late Triassic	Hosselkus			Shasta
Ichthyosauria	Shastasauridae	Toretocnemus	perrini	UCMP	9118		Late Triassic	Hosselkus			Shasta
Ichthyosauria				UCMP	9018		Late Triassic	Hosselkus			Shasta
Ichthyosauria				UCMP	10632		Late Triassic	Hosselkus	UC party		Shasta
Ichthyosauria				UCMP	10633		Late Triassic	Hosselkus	UC party		Shasta
Ichthyosauria				UCMP	10639		Late Triassic	Hosselkus	UC party		Shasta
Ichthyosauria				UCMP	9153		Late Triassic	Hosselkus		1904	Shasta
Ichthyosauria				SC	VRD21	Bone?	Late Triassic	Hosselkus	Hilton, R.	1973	Shasta
Ichthyosauria				UCMP	94502	Skull (incompl.)	Late Triassic	Hosselkus	UC party	1901	Shasta
Ichthyosauria				LACM	56751	Vertebra frag.	Late Triassic	Hosselkus	Wilson, E. C.	1973	Shasta
Ichthyosauria				SC	VRD20	Vertebrae	Late Triassic	Hosselkus	Hilton, R.	1973	Shasta
Ichthyosauria?				SC	VR17	Fragment	Late Triassic	Hosselkus	Hilton, R.	1973	Shasta
Ichthyosauria?				SC	VR5	Fragment	Late Triassic	Hosselkus	Hilton, R.	1973	Shasta
Ichthyosauria?				SC	VR18	Vertebrae	Late Triassic	Hosselkus	Hilton, R.	1973	Shasta
				UCMP	9088	Skeleton	Late Triassic	Hosselkus			Shasta
				UCMP	9615		Late Triassic	Hosselkus		1903	Shasta
				UCMP	10777	Cranial bone?	Late Triassic	Hosselkus	Merriam, J. C.	1906	Shasta
				UCMP	10790	Limb bone	Late Triassic	Hosselkus	Merriam, J. C.	1906	Shasta
				UCMP	1063		Late Triassic	Hosselkus			Shasta
				UCMP	10630		Late Triassic	Hosselkus	UC party		Shasta
				UCMP	1064		Late Triassic	Hosselkus			Shasta
				UCMP	10090		Late Triassic	Hosselkus	Furlong, E. L.	1904	Shasta
				UCMP	10641	Rib?	Late Triassic	Hosselkus			Shasta
				UCMP	10804		Late Triassic	Hosselkus	Merriam, J. C.	1906	Shasta
				UCMP	10805	Bone frags.	Late Triassic	Hosselkus	Merriam, J. C.	1906	Shasta
				UCMP	10806		Late Triassic	Hosselkus	Merriam, J. C.	1906	Shasta
				UCMP	10807	Limb bones	Late Triassic	Hosselkus	Merriam, J. C.	1906	Shasta
				UCMP	10808	Bone	Late Triassic	Hosselkus	Merriam, J. C.	1906	Shasta
				UCMP	10812	Bone	Late Triassic	Hosselkus	Merriam, J. C.	1906	Shasta
				UCMP	10819		Late Triassic	Hosselkus	Merriam, J. C.	1906	Shasta
				UCMP	19477		Late Triassic	Hosselkus	UC party	1903	Shasta
				UCMP	9094		Late Triassic	Hosselkus	Alexander, A. M. & Furlong, E. L.	1903	Shasta
				UCMP	9611		Late Triassic	Hosselkus			Shasta
				UCMP	9612		Late Triassic	Hosselkus			Shasta
				UCMP	9613		Late Triassic	Hosselkus			Shasta
				UCMP	10530	Vertebra	Late Triassic	Hosselkus	UC party	1909	Shasta
				UCMP	10754		Late Triassic	Hosselkus	Merriam, J. C.	1906	Shasta

GROUP	TYPE	GENUS	SPECIES	REPOSITORY	SPECIMEN NO.	ELEMENTS	AGE	FORMATION	FINDER	FIND DATE	PROVINCE
				UCMP	10761		Late Triassic	Hosselkus	UC party	1906	Shasta
				UCMP	10762		Late Triassic	Hosselkus	UC party	1906	Shasta
				UCMP	10763		Late Triassic	Hosselkus	UC party	1906	Shasta
				UCMP	10785		Late Triassic	Hosselkus	Merriam, J. C.	1906	Shasta
				UCMP	10786		Late Triassic	Hosselkus	Merriam, J. C.	1906	Shasta
				UCMP	10789		Late Triassic	Hosselkus	Merriam, J. C.	1906	Shasta
				UCMP	10795		Late Triassic	Hosselkus	Merriam, J. C.	1906	Shasta
				UCMP	10796		Late Triassic	Hosselkus	Merriam, J. C.	1906	Shasta
				UCMP	10798	Vertebra, neural arch + limb bone	Late Triassic	Hosselkus	Merriam, J. C.	1906	Shasta
				UCMP	10810		Late Triassic	Hosselkus	Merriam, J. C.	1906	Shasta
				UCMP	10813		Late Triassic	Hosselkus	Merriam, J. C.	1906	Shasta
				UCMP	10820		Late Triassic	Hosselkus	Merriam, J. C.	1906	Shasta

CALIFORNIA JURASSIC MARINE REPTILES

Plesiosauria

GROUP	TYPE	GENUS	SPECIES	REPOSITORY	SPECIMEN NO.	ELEMENTS	AGE	FORMATION	FINDER	FIND DATE	PROVINCE
Plesiosauria	?Cryptoclididae			UCMP	44696	Skeleton (incompl.)	Late Jur.	Mariposa, Salt Spring Member	Clark, L. & Welles, S.	1954	Mariposa
Plesiosauria	Elasmosauridae			LACM	56073	Vertebra cervical	Jurassic?	Bedford Canyon	Kanakoff, G. P.		Orange
Plesiosauria	Plesiosauridae	Plesiosaurus	hesternus	UCMP	41599	Vertebra	Late Jur.	Franciscan	Olsonowski, B. & Durham, J. W.	1949	San Luis Obispo
Plesiosauria	Plesiosauridae			SC	VRS9	Vertebrae	Late Jur.	Great Valley Group	Smith, J.	1902	Glenn
Plesiosauria	Plesiosauridae			SC		Humerus, lft. distal end	Late Jur.	Great Valley Group	Kirkland, K.	1999	Glenn
Plesiosauria	Plesiosauridae			SC	VR77	Rib, dorsal, proximal end	Late Jur.	Great Valley Group	Maitia, D.	1999	Tehama
Plesiosauria	Plesiosauridae			SC	VR78	Gastralia	Late Jur.	Great Valley Group	Hilton, P.	1999	Tehama
Plesiosauria	Plesiosauridae			SC	VRS1	Vertebrae	Late Jur.	Great Valley Group	Kelly, I.	ca. 1950	Tehama
Plesiosauria	Plesiosauridae			SC	VRS10	Tarsal cf.	Late Jur.	Great Valley Group	Maitia, D.	1999	Tehama
Plesiosauria	Plesiosauridae			SC	VRS6	Neural arch	Late Jur.	Great Valley Group	Jensen, J., Jr.	1999	Tehama
Plesiosauria	Plesiosauridae			SC	VRS8	Ischium, lft.?	Late Jur.	Great Valley Group	Hilton, R.	1999	Tehama
Plesiosauria	Plesiosauridae?			SC	VRS11	Rib?, distal end	Late Jur.	Great Valley Group	Maitia, D.	1999	Tehama?
Plesiosauria	Pliosauridae	Dolichorhynchops (trinacromerum)		UCMP	56204	Vertebra cervical, metatarsal? + frag.	Late Jur.	Great Valley Group	McKee, B.	1954	Tehama
Plesiosauria				UCMP	56204	Vertebra cervical (posterior)	Late Jur.	Great Valley Group	Wahl-Hall, D. R. & McKee, E. D.	1959	Tehama
Plesiosauria				SC	VRS5	Neural arch	Late Jur.	Great Valley Group	Jensen, J., Jr.	1999	Tehama
Plesiosauria				SC	VRS7	Coricoid	Late Jur.	Great Valley Group	Hilton, R.	1999	Tehama
Plesiosauria				lost		Bone	Late Jur.	Great Valley Group	McKee, B.	1960	Tehama
Plesiosauria				lost		Vertebrae + ribs	Late Jur.	Great Valley Group	Knock, T. L.	1913	Glenn
Plesiosauria				BC		Vertebrae	Late Jur.	Great Valley Group	Mattison, D.		Tehama

GROUP	TYPE	GENUS	SPECIES	REPOSITORY	SPECIMEN NO.	ELEMENTS	AGE	FORMATION	FINDER	FIND DATE	PROVINCE
Plesiosauria				SC	VRS3	Vertebrae	Late Jur.	Great Valley Group	Hilton, R.	1998	Tehama
Plesiosauria				SC	VRS4	Coricoid	Late Jur.	Great Valley Group	Maitia, D.	1998	Tehama
Plesiosauria				SC	VRS13	Tooth	Late Jur.	Great Valley Group	Bennison, A.	2000	Glenn
Ichthyosaurs											
Ichthyosauria	Ichthyosauridae	*Ichthyosaurus*	*franciscanus*	UCMP	33432	Rostrum	Late Jur.	Franciscan	Smith, N. J.	1935	San Joaquin
Ichthyosauria	Ichthyosauridae	*Ichthyosaurus*	*californicus*	UCMP	36394	Skull (anterior)	Late Jur.	Franciscan	Hammond, J.	1940	Stanislaus
Misc. reptiles											
Reptilia				SC	VR70	Bone frag.	Late Jur.	Great Valley Group	Hilton, R.	1998	Tehama
Reptilia				SC	VR74	Fragment	Late Jur.	Great Valley Group	Hilton, R.	1998	Tehama
Reptilia				SC	VR75	Fragment	Late Jur.	Great Valley Group	Jensen, J., Jr.	1998	Tehama
Reptilia				SC	VR79	Fragment	Late Jur.	Great Valley Group	Hilton, R.	1998	Tehama
Reptilia				UCMP	114156	Fragment	Late Jur.	Knoxville	Young, G.	1957	Tehama
California Cretaceous birds											
bird		*Ichthyornis*		UCMP	170785	Humerus	Late Cret.	Chico	Göhre, E.	1998	Butte
bird	Neognathae			UCMP	171185	Ulna, lft.	Late Cret.	Chico	Göhre, E.	1998	Butte
bird		*Hesperornis*		SC	VBHE1	Phalange	Late Cret.	Chico	Göhre, E.	2000	Butte
CALIFORNIA CRETACEOUS MARINE REPTILES											
Pterosaurs											
Pterosauria				SC	VRF8	Fragment	Early Cret.	Budden Canyon	Embree, P.	1998	Shasta
Pterosauria				SC	VRF5	Metacarpal (4th)	Late Cret.	Chico	Göhre, E.	1998	Butte
Pterosauria				SC	VRF6	Ulna	Late Cret.	Chico	Göhre, E.	1999	Butte
Pterosauria				SC	VRF7	Finger frag.	Late Cret.	Chico	Hilton, R.	1999	Butte
Ichthyosaur											
Ichthyosaur	"Shastasauridae"			LACM	27798	Skull frags. + teeth	{Ku}, "Rhae." *(sic)*	{Great Valley Group} Hosselkus *(sic)*		after 1936	Kern
Mosasaurs											
Squamata	Mosasauridae	*Plesiotylosaurus*	*crassidens*	AMNH	1490		Late Cret.	Moreno			Fresno
Squamata	Mosasauridae	*Plesiotylosaurus*	*crassidens*	LACM/CIT	2750	Skeleton w/ skull + mandibles	Late Cret.	Moreno	Leard, R. M.	1939	Fresno
Squamata	Mosasauridae	*Plesiotylosaurus*	*crassidens*	LACM/CIT	2753	Skull	Late Cret.	Moreno	Leard, R. M., et al.	1939	Fresno
Squamata	Mosasauridae	*Plesiotylosaurus*	*crassidens*	LACM/CIT	2759	Skull + mandibles	Late Cret.	Moreno	Leard, R. M.	1939	Fresno
Squamata	Mosasauridae	*Plesiotylosaurus*	*crassidens*	UCMP	126716	Skeleton (incompl.)	Late Cret.	Moreno	Yang-Staebler, D.	1981	Fresno
Squamata	Mosasauridae	*Plesiotylosaurus*	*crassidens*	UCMP	137249	Skeleton (incompl.)	Late Cret.	Moreno	Staebler, C.	1980	Fresno
Squamata	Mosasauridae	*Plesiotylosaurus*	*crassidens*	UCMP	137248	Skeleton (incompl.)	Late Cret.	Moreno	Desatoff, P.	1981	Fresno
Squamata	Mosasauridae	*Plotosaurus (Kolposaurus)*	*bennisoni*	UCMP	32778	Skull, dentary + vertebrae (anterior)	Late Cret.	Moreno	Bennison, A. & Hesse, S.	1937	Stanislaus
Squamata	Mosasauridae	*Plotosaurus*	*tuckeri*	LACM/CIT	2751	Vertebra caudal	Late Cret.	Moreno	Leard, R. M., et al.	1939	Fresno

GROUP	TYPE	GENUS	SPECIES	REPOSITORY	SPECIMEN NO.	ELEMENTS	AGE	FORMATION	FINDER	FIND DATE	PROVINCE
Squamata	Mosasauridae	Plotosaurus	tuckeri	LACM/CIT	2755	Vertebra caudal	Late Cret.	Moreno	Leard, R. M., et al.	1939	Fresno
Squamata	Mosasauridae	Plotosaurus	tuckeri	UCMP	45299	Vertebra caudal	Late Cret.	Moreno	Welles party	1956	Fresno
Squamata	Mosasauridae	Plotosaurus	tuckeri	UCMP	45301	Vertebra	Late Cret.	Moreno	Welles party	1956	Fresno
Squamata	Mosasauridae	Plotosaurus	tuckeri	UCMP	32943	Vertebra caudal (3)	Late Cret.	Moreno	Thompson, M. M. & Bennison, A.	1936	Stanislaus
Squamata	Mosasauridae	Plotosaurus (Kolp.)	tuckeri	UCMP	33913	Vertebrae	Late Cret.	Moreno	UC party	1937	Fresno
Squamata	Mosasauridae	Plotosaurus		LACM/CIT	2756	Vertebra caudal	Late Cret.	Moreno	Leard, R. M., et al.	1939	Fresno
Squamata	Mosasauridae	Plotosaurus		LACM/CIT	2757	Vertebra	Late Cret.	Moreno	Leard, R. M., et al.	1939	Fresno
Squamata	Mosasauridae	Plotosaurus		UCMP	36050	Vertebrae caudal (2)	Late Cret.	Moreno	Walker, H. G.	1920	Fresno
Squamata	Mosasauridae	Plotosaurus		UCMP	45302	Vertebral column (incompl.)	Late Cret.	Moreno	Sheldon	1956	Fresno
Squamata	Mosasauridae	Plotosaurus		UCMP	45303	Skull + vertebra	Late Cret.	Moreno	Hosley	1956	Fresno
Squamata	Mosasauridae	Plotosaurus		UCMP	57671	Vertebra	Late Cret.	Moreno	(Cosgriff, J.) UC party	1956	Fresno
Squamata	Mosasauridae	Plotosaurus		UCMP	57672	Vertebra	Late Cret.	Moreno	West or Hildebrant, H. & UC party	1960	Fresno
Squamata	Mosasauridae	Plotosaurus		UCMP	57673	Vertebra caudal	Late Cret.	Moreno	UC party	1957	Fresno
Squamata	Mosasauridae	Plotosaurus		UCMP	57674		Late Cret.	Moreno	UC party	1957	Fresno
Squamata	Mosasauridae	Plotosaurus		UCMP	57675	Vertebrae caudal	Late Cret.	Moreno	UC party	1957	Fresno
Squamata	Mosasauridae	Plotosaurus		UCMP	57676	Vertebra	Late Cret.	Moreno	UC party	1956	Fresno
Squamata	Mosasauridae	Plotosaurus		UCMP	126284	Vertebrae, pelvic frags. + pes	Late Cret.	Moreno	Staebler, C. N. & Dake, P.	1980	Fresno
Squamata	Mosasauridae	Plotosaurus		UCMP	138214	Vertebrae	Late Cret.	Moreno	Hutchison, J. H.	1992	Fresno
Squamata	Mosasauridae	Plotosaurus				Paddle w/ claw	Late Cret.	Moreno	Staebler, A. E.	1985	Fresno
Squamata	Mosasauridae	Plotosaurus?				Skeleton (incompl.)	Late Cret.	Moreno	Staebler, C. N.	1984	Fresno
Squamata	Mosasauridae			Göhre	private	Vertebra	Late Cret.	Chico	Göhre, E.	1998	Butte
Squamata	Mosasauridae			LACM	52685	Skull frags.	Late Cret.	Moreno			Fresno
Squamata	Mosasauridae			LACM	52686	Phalanx	Late Cret.	Moreno			Fresno
Squamata	Mosasauridae			LACM/CIT	2945	Skull w/ mandibles	Late Cret.	Moreno	Leard, R. M., et al.	1940	Fresno
Squamata	Mosasauridae			MPC		Skeleton (baby?)	Late Cret.	Moreno	Howard, D.	1959	Fresno
Squamata	Mosasauridae			private		Teeth (3)	Late Cret.	Moreno	Kepper, J.	ca. 1950	Fresno
Squamata	Mosasauridae	Clidastes?		SC	VR59	Skull frags.	Late Cret.	Chico	Antuzzi, P.	1994	Placer
Squamata	Mosasauridae			SDNHM	25895	Vertebrae caudal (2)	Late Cret.	Pt. Loma	Riney, B.	1982	San Diego
Squamata	Mosasauridae			SDNHM	67369	Vertebrae w/ ribs	Late Cret.	Pt. Loma	Cerrutti, R.	1986	San Diego
Squamata	Mosasauridae			SDNHM			Late Cret.	Pt. Loma	Riney, B.		San Diego
Squamata	Mosasauridae			Stanford		Vertebrae	Late Cret.	Moreno	Fonseca, C.	1997	Fresno
Squamata	Mosasauridae			UCMP	45300	Vertebra	Late Cret.	Moreno	Welles party	1956	Fresno
Squamata	Mosasauridae			UCMP	57582	Skull w/ cervical vertebrae to pectrum	Late Cret.	Moreno	UC party	1958	Fresno
Squamata	Mosasauridae			UCMP	65101	Vertebra caudal	Late Cret.	Moreno	Webb, S. D.	1963	Fresno

GROUP	TYPE	GENUS	SPECIES	REPOSITORY	SPECIMEN NO.	ELEMENTS	AGE	FORMATION	FINDER	FIND DATE	PROVINCE
Squamata	Mosasauridae			UCMP	79687	Tooth	Late Cret.	Moreno	Welles party	1957	Fresno
Squamata	Mosasauridae			UCMP	125973	Vertebra caudal w/ chevrons	Late Cret.	Moreno	Staebler, C. N.	1980	Fresno
Squamata	Mosasauridae			UCMP	126277	Vertebral column (incompl.)	Late Cret.	Moreno	Staebler, C. N. & Staebler, A. E.	1975	Fresno
Squamata	Mosasauridae			UCMP	126278	Skeleton (incompl.)	Late Cret.	Moreno	Staebler, C. N.	1982	Fresno
Squamata	Mosasauridae			UCMP	126279	Skeleton (incompl., postcranial)	Late Cret.	Moreno	Staebler, C. N.	1976	Fresno
Squamata	Mosasauridae			UCMP	126280	Skull (posterior), lower mandible frags. +	Late Cret.	Moreno	Staebler, C. N.	1980	Fresno
Squamata	Mosasauridae			UCMP	126281	Vertebrae caudal	Late Cret.	Moreno	Staebler, C. N.	1980	Fresno
Squamata	Mosasauridae			UCMP	126282	Skeleton (incompl.)	Late Cret.	Moreno	Staebler, C. N.	1980	Fresno
Squamata	Mosasauridae			UCMP	126283	Skull + pectoral bones	Late Cret.	Moreno	Staebler, C. N.	1980	Fresno
Squamata	Mosasauridae			UCMP	126715	Skeleton (incompl.)	Late Cret.	Moreno	Staebler, C. N.	1982	Fresno
Squamata	Mosasauridae			UCMP	137245	Skeleton (incompl.)	Late Cret.	Moreno	Desatoff, P.	1981	Fresno
Squamata	Mosasauridae			UCMP	137246	Skeleton (incompl.)	Late Cret.	Moreno	Staebler, C. N.	1982	Fresno
Squamata	Mosasauridae			UCMP	137247	Skeleton (incompl.)	Late Cret.	Moreno	Staebler, C. N.	1981	Fresno
Squamata	Mosasauridae			UCMP	137250	Skeleton (incompl.)	Late Cret.	Moreno	Goodwin, M.	1980	Fresno
Squamata	Mosasauridae			UCMP	137250	Mandible	Late Cret.	Moreno	Staebler, C. N.		Fresno
Squamata	Mosasauridae			UCMP	137251	Skeleton (incompl.)	Late Cret.	Moreno	Staebler, C. N.	1980	Fresno
Squamata	Mosasauridae			UCMP	140276	Mandible, scapula + humerus	Late Cret.	Moreno			Fresno
Squamata	Mosasauridae			UCMP	acc4309		Late Cret.	Moreno	Staebler, C. N.	1991	Fresno
Squamata	Mosasauridae			UCMP	acc4309		Late Cret.	Moreno	Staebler, C. N.	1980	Fresno
Squamata	Mosasauridae			UCMP	acc4309		Late Cret.	Moreno	Staebler, C. N.	1979	Fresno
Squamata	Mosasauridae			UCMP	acc4309		Late Cret.	Moreno	Staebler, C. N.	1980	Fresno
Squamata	Mosasauridae			UCMP	acc4309		Late Cret.	Moreno	Staebler, C. N.	1991	Fresno
Squamata	Mosasauridae			UCMP	acc4309		Late Cret.	Moreno	Staebler, C. N.	1985	Fresno
Squamata	Mosasauridae			UCMP	acc4309		Late Cret.	Moreno	Staebler, C. N.	1988	Fresno
Squamata	Mosasauridae			UCMP	acc4309		Late Cret.	Moreno	Staebler, C. N.	1988	Fresno
Squamata	Mosasauridae			UCMP	acc4309		Late Cret.	Moreno	Staebler, C. N.	1987	Fresno
Squamata	Mosasauridae			UCMP	acc4309		Late Cret.	Moreno	Staebler, C. N.	1985	Fresno
Squamata	Mosasauridae			UCMP	acc4309		Late Cret.	Moreno	Staebler, C. N.	1984	Fresno
Squamata	Mosasauridae			UCMP	acc4309		Late Cret.	Moreno	Staebler, C. N.	1984	Fresno
Squamata	Mosasauridae			UCMP	acc4309		Late Cret.	Moreno	Staebler, C. N.	1980 or 1982	Fresno
Squamata	Mosasauridae			UCMP	acc4309		Late Cret.	Moreno	Staebler, C. N.	1982	Fresno
Squamata	Mosasauridae			UCMP	acc4309		Late Cret.	Moreno	Yang, D.	1981	Fresno
Squamata	Mosasauridae			UCMP	acc4309		Late Cret.	Moreno	Staebler, C. N.	1981	Fresno
Squamata	Mosasauridae			UCMP	acc4309		Late Cret.	Moreno	Staebler, C. N.	1981	Fresno

GROUP	TYPE	GENUS	SPECIES	REPOSITORY	SPECIMEN NO.	ELEMENTS	AGE	FORMATION	FINDER	FIND DATE	PROVINCE
Squamata	Mosasauridae			UCMP	acc4309		Late Cret.	Moreno	Staebler, C. N.	1981	Fresno
Squamata	Mosasauridae			UCMP	acc4309		Late Cret.	Moreno		1981	Fresno
Squamata	Mosasauridae			UCMP	acc4309		Late Cret.	Moreno	Staebler, C. N.	1981	Fresno
Squamata	Mosasauridae			UCMP	acc4309		Late Cret.	Moreno	Staebler, C. N.	1981	Fresno
Squamata	Mosasauridae			UCMP	acc4309		Late Cret.	Moreno	Staebler, C. N.	1980	Fresno
Squamata	Mosasauridae			UCMP	acc4309		Late Cret.	Moreno	Staebler, C. N.	1980	Fresno
Squamata	Mosasauridae			UCMP	acc4309		Late Cret.	Moreno	Staebler, C. N.	1980	Fresno
Squamata	Mosasauridae			UCMP	acc4309		Late Cret.	Moreno	Staebler, C. N.	1980	Fresno
Squamata	Mosasauridae			UCMP	acc4309		Late Cret.	Moreno	Staebler, C. N.	1980	Fresno
Squamata	Mosasauridae			UCMP		Vertebrae caudal?	Late Cret.	Moreno	Staebler, C. N.	1987	Fresno
Squamata	Mosasauridae			UCMP		Vertebrae caudal?	Late Cret.	Moreno	Staebler, C. N.	1987	Fresno
Squamata	Mosasauridae					Pelvic paddle, lft.	Late Cret.	Moreno	Dake, P.	1982	Butte
Squamata	Mosasauridae?			SC	VR68	Tooth	Late Cret.	Chico	Göhre, E.	1998	Butte
Squamata				UCMP	131812	Bone frags.	Late Cret.	Moreno	Staebler, C. N.	1980	Fresno
Plesiosauria											
Plesiosauria	Elasmosauridae	Hydrotherosaurus	alexandrae	UCMP	33912	Skeleton	Late Cret.	Moreno	Paiva, F./Fresno State & UC	1937	Fresno
Plesiosauria	Elasmosauridae	Morenosaurus	stocki	UCMP	40172	Skull	Late Cret.	Moreno			Fresno
Plesiosauria	Elasmosauridae	Morenosaurus	stocki	UCMP	66993	Vertebrae caudal	Late Cret.	Moreno	Zarconi, C. & Lovenberg, M.	1964	Merced
Plesiosauria	Elasmosauridae			LACM/CIT	2754	Limb bones + vertebrae	Late Cret.	Moreno	Leard, R. M.	1939	Fresno
Plesiosauria	Elasmosauridae			LACM	52687	Vertebra	Late Cret.	Moreno	Wahrhaftig, C.?	1939?	Fresno
Plesiosauria	Elasmosauridae			UCMP	79688	Vertebra	Late Cret.	Moreno	Welles party	1957	Fresno
Plesiosauria	Elasmosauridae			UCMP	139287	Skeleton	Late Cret.	Moreno	Staebler, C. N.	1989	Fresno
Plesiosauria	Elasmosauridae			SBCM	17870	Centrum	Late Cret.	San Francisquito?	Colman, R.	1978	San Bernardino
Plesiosauria	Elasmosauridae			SBCM	A500-6872	Vertebrae (40)	Late Cret.	San Francisquito?			San Bernardino
Plesiosauria	Plesiosauridae			UCMP	57581	Neck + paddle	Late Cret.	Moreno	UC party	1957	Fresno
Plesiosauria	Plesiosauridae			UCMP	69688	Vertebrae	Late Cret.	Moreno	Howard, D.	1957	Fresno
Plesiosauria	Plesiosauridae			LACM/CIT	33396	Vertebra	Late Cret.	Moreno	Ray, J.	1954	Fresno
Plesiosauria				UCMP?		Tooth	Late Cret.	"Chico"		1908	Fresno
Plesiosauria				UCMP	acc4309	Vertebrae?	Late Cret.	Moreno	Horner, J. & Staebler, C. N.	1988	Fresno
Plesiosauria				UCMP	43588	Centrum	Late Cret.	Ladd	Kirk, M. V.	1950	Orange
Plesiosauria				UCMP	45298	Vertebra	Late Cret.	Moreno	Welles party	1956	San Benito
Plesiosauria		Aphrosaurus	furlongi	LACM/CIT	2832	Skeleton (anterior)	Late Cret.	Moreno	White, R. M.	1939	Fresno
Plesiosauria		Aphrosaurus	furlongi	LACM/CIT	2748	Gastroliths	Late Cret.	Moreno	Stock, C.		Fresno
Plesiosauria		Aphrosaurus	furlongi	LACM/CIT	2748	Limb bones, vertebrae + ribs	Late Cret.	Moreno	Leard, R. M., et al.	1939	Fresno
Plesiosauria		Aphrosaurus	furlongi	YPM-PU	16432/ CIT2748	Gastroliths	Late Cret.	Moreno	Stock, C.		Fresno

GROUP	TYPE	GENUS	SPECIES	REPOSITORY	SPECIMEN NO.	ELEMENTS	AGE	FORMATION	FINDER	FIND DATE	PROVINCE
Plesiosauria		*Fresnosaurus*	*drescheri*	LACM/CIT	2758	Limb bones	Late Cret.	Moreno	Leard, R. M., et al.	1939	Fresno
Plesiosauria		*Morenosaurus*	*stocki*	LACM/CIT	2749	Skull, mandibles + limb bones	Late Cret.	Moreno	Leard, R. M., et al.	1939	Fresno
Plesiosauria		*Morenosaurus*	*stocki*	LACM/CIT	2802	Skeleton	Late Cret.	Moreno	Drescher, A. B. & Leard, R. M.	1940	Fresno
Plesiosauria				Göhre	private	Tooth	Late Cret.	Chico	Embree, P.	1999	Butte
Turtles											
Cheloniid	Adocidae	*Adocus (Basilemys)*		UCMP	126717	Skull, mandibles + carapace (anterior) +	Late Cret.	Moreno	Staebler, C. N.	1979	Fresno
Cheloniid	Adocidae	*Basilemys*		ETC = OCPC	GC958122	15 carapace frags.	Late Cret.	Ladd	Calvano, G.		Orange
Cheloniid	Cheloniidae	"*Toxochelyid*"		SC	VRT32	Supraoccipital	Late Cret.	Chico	Hilton, R.	1996	Placer
Cheloniid	Dermochelyidae			UCMP	172070	Scapula, lft.	Late Cret.	Chico	Göhre, E.	1998	Butte
Cheloniid	Dermochelyidae			UCMP	172071	Humerus, lft.	Late Cret.	Chico	Göhre, E.	1998	Butte
Cheloniid	Dermochelyidae			SC	VRT19	Coracoid, lft.	Late Cret.	Chico	Antuzzi, P.	1995	Placer
Cheloniid	Toxochelyidae	*Osteopygis*		UCMP	123616	Skull, mandibles + rt. and lft. humerus +	Late Cret.	Moreno	Staebler, C. N.	1979	Fresno
Cheloniid				ETC = OCPC	GC9781174 = 22183, 22184	Costal, scute	Late Cret.	Williams	Calvano, G.	1997	Orange
Cheloniid				ETC = OCPC	GC9781766	Costal			Calvano, G.	1997	Orange
Cheloniid				LACM/CIT	3228	Carapace, vertebrae + limb bones	Late Cret.	Moreno	Smith, B.	1941	Fresno
Cheloniid				SC	VR62	Costal frag.	Late Cret.	Chico	Antuzzi, P.	1996	Placer
Cheloniid				SC	VRT23	Scapula (incompl.)	Late Cret.	Chico	Antuzzi, P.	1994	Placer
Cheloniid				SC	VRT24	Femur	Late Cret.	Chico	Antuzzi, P.	1994	Placer
Cheloniid				SC	VRT25	Limb frag.	Late Cret.	Chico	Hilton, R.	1994	Placer
Cheloniid				SC	VRT26	Femur?, proximal end	Late Cret.	Chico	Antuzzi, P.	1995	Placer
Cheloniid				SC	VRT27	Costal frag.	Late Cret.	Chico	Hilton, R.	1996	Placer
Cheloniid				SC	VRT31		Late Cret.	Chico	Hilton, R.	1998	Butte
Cheloniid				SDNHM	126276	Carapace frag.	Late Cret.	Cabrillo	Riney, B.	late 1980s	San Diego
Cheloniid				UCMP	126276	Plastron frag.	Late Cret.	Moreno	Moise, W.	1980	Fresno
Cheloniid				UCMP	136419	Costal, proximal end	Late Cret.	Chico	Hilton, R.	1988	Tehama
Cheloniid				UCMP	169162	Supraoccipital (base)	Late Cret.				
Cheloniid				UCMP	172072	Costal frag.	Late Cret.	Chico	Göhre, E.	1998	Butte
Cheloniid				UCMP	172073	Periferal	Late Cret.	Chico	Göhre, E.	1998	Butte
Cheloniid				UCMP	172074	Fragment	Late Cret.	Chico	Göhre, E.	1998	Butte
Cheloniid				UCMP	172075	Maxilla frag.	Late Cret.	Chico	Göhre, E.	1998	Butte
Cheloniid				UCMP		Skull endocast	Late Cret.	Moreno	Staebler, C. N.	1980	Fresno
Misc. reptiles											
Reptilia				UCMP	70988	Tooth	Late Cret.	Panoche	Peck, Nelson, Allison & Swan	1964	Contra Costa

GROUP	TYPE	GENUS	SPECIES	REPOSITORY	SPECIMEN NO.	ELEMENTS	AGE	FORMATION	FINDER	FIND DATE	PROVINCE
Reptilia				LACM/CIT	2752	Vertebra	Late Cret.	Moreno	Leard, R. M., et al.	1939	Fresno
Reptilia				LACM/CIT	2804	Vertebra caudal	Late Cret.	Moreno	Drescher, A. B. & Leard, R. M.	1940	Fresno
Reptilia				LACM	120441	Vertebra	Late Cret.	Moreno	Caltech party	1939	Fresno
Reptilia				LACM	120477	Tooth	Late Cret.	Moreno	Caltech party		Fresno
Reptilia				CSUF		Shoulder girdle w/ vertebrae	Late Cret.	Moreno			Fresno
Reptilia				UCMP	36252	Tooth	Late Cret.	Quinto	Bennison, A.	1940	Merced
Reptilia				SC	VR56	Fragment	Late Cret.	Chico	Hilton, R.	1996	Placer
Reptilia?				SC	VR60	Fragment	Late Cret.	Chico	Antuzzi, P.	1996	Placer
Reptilia				SC	VR61	Fragment w/ tooth impression	Late Cret.	Chico	Antuzzi, P.	1996	Placer
Reptilia				SC	VR64	Bone	Late Cret.	Chico	Hilton, R.	1996	Placer
Reptilia				SC	VR65	Bone	Late Cret.	Chico	Antuzzi, P.	1996	Placer
Reptilia				CAS	159	Fragment	Early Cret.	Budden Canyon	Shouckman, C.		Shasta
Reptilia				SC	VR**51	Tibia	Late Cret.	Chico	Antuzzi, P.	1994	Placer
Reptilia				ETC = OCPC	GC9781.174 = 22183, 22184	Fragments	Late Cret.	Williams	Peck, P.	1996	Orange

GLOSSARY

ACCRETE: To add new crustal material to a continent or island arc through the actions of plate tectonics.

AMMONITE: A shelled cephalopod mollusk (relative of the chambered nautilus, squid, and octopus) that lived in the Mesozoic.

AMMONOID: Referring to a group of cephalopod mollusks that includes the ammonites and other shelled cephalopods.

ANAPSID: A subclass of reptiles that has no openings in its skull behind the orbit (eye socket).

ANKYLOSAUR: A group of medium-sized, highly armored, herbivorous reptiles that possessed a gnarly tail club.

AQUATIC: Living in a water environment (usually freshwater).

ARCHOSAUR: A subgroup of the diapsids that gave rise to the thecodonts (the order of reptiles whose teeth insert into sockets).

ARTICULATED: Referring to bones that are found arranged as they were when the animal was living, rather than being separate from each other.

AVIFAUNA: Birds.

BASAL TETRAPOD: A primitive vertebrate having four legs.

BASALT(IC): Pertaining to dark volcanic rocks rich in iron and magnesium.

BELEMNITE: The internal skeleton of a squidlike cephalopod found primarily in Jurassic and Cretaceous marine sedimentary rocks.

BIPEDAL: Walking on two legs.

BRACKISH: Describing water intermediate in salinity between seawater and freshwater.

CARAPACE: A bony or chitinous case covering the dorsal (upper portion) of an animal. In a turtle this would be the upper "shell."

CARNOSAUR: A meat-eating dinosaur; specifically, one of the big-headed theropods with powerful legs and small arms, such as *Allosaurus.*

CAUDAL: Referring to the tail region or tail bones of a vertebrate.

CEPHALOPOD: Literally meaning "head-foot," a member of a highly evolved group of mollusks having a head surrounded by tentacles. Most have two eyes and may use jet propulsion. Many fossil forms had chambered shells, as does today's chambered nautilus.

CERATOPSIAN: Referring to a group of ornithischian dinosaurs that typically had a beaklike snout and flaring jugals (cheekbones). Like *Triceratops,* they often had armored skulls containing horns and frills.

CLADE: A group of organisms that are genetically related (monophyletic group).

CLADISTIC: Having to do with the grouping of organisms according to monophyletic systematics.

CLASTIC: Referring to rock or sediment that is made out of pieces of other rocks and/or minerals.

COELUROSAUR: Smaller and more nimble than carnosaurs, coelurosaurs were theropods with long, narrow jaws and longer, grasping arms.

CONCRETION: A harder or more compact, usually rounded portion of a better-cemented part of a sedimentary (sometimes pyroclastic) rock. Concretions often surround a nucleus such as a fossil bone, shell, wood, or leaf.

CONGLOMERATE: A sedimentary rock made out of rounded pebbles, cobbles, or boulders cemented by a matrix of finer clastic materials and minerals.

COPROLITE: Fossil feces.

CORACOID: A bone from the ventral side of the shoulder girdle that joins with the scapula.

CRUST: The outer portion of the solid earth.

CYCAD: A cone-bearing, palmlike gymnosperm, often referred to as a sego palm by commercial nurseries.

DEPOSITIONAL SETTING: A place, often under water, where the energy of moving, sediment-bearing water slows down; hence sediment rains out and is often deposited in layers.

DERMAL: Pertaining to skin.

DIAPSID: A subclass of reptiles possessing two holes behind each orbit (eye socket) of the skull.

DISTAL: Pertaining to the portion of something (such as the end of a bone) that is away from the body rather than the portion that is closer to it (proximal).

DORSAL FIN: The fin that is on or near the mid-back of a swimming vertebrate.

DROMAEOSAURID: A coelurosaur with a large raptorial claw on the second toe of the hind foot.

ECHINODERM: Literally meaning "spiny skin," echinoderms represent a phylum of invertebrates with five-part radial symmetry such as urchins, sea lilies, and sea stars.

ECHOLOCATION: The process of using sound to locate one's surroundings, such as when submarines use sonar.

ELASMOSAURID: A broadly used term for long-necked plesiosaurs.

ERA: One of the major units of geologic time, as in the Mesozoic Era.

FLYSCH: A series of marine sedimentary rocks, usually deposited offshore, containing abundant fine-grained clastic strata that are somewhat rhythmically punctuated by coarser layers.

FOREARC BASIN: A sedimentary basin that is seaward of an arc (often volcanic) of mountains.

FORMATION: A fundamental unit of rock strata that consists of a particular set of rocks (and perhaps fossils) with lateral, vertical, and temporal constraints.

GASTRALIA: Belly ribs that act as armor and help to support the viscera.

GASTROLITH: Literally meaning "stomach stone," these rounded and smoothed stones are found within the gut region of a variety of reptiles and birds. They are thought to be ingested for processing food and/or for ballast.

GENUS: A group name used to classify organisms, usually consisting of one or more smaller divisions (species) and belonging to a larger group called a family.

GINKGO: The common name for an order of gymnosperms that flourished during the Mesozoic but today is represented by only a single species.

GRANITIC: A loose reference to coarse-grained, intrusive (as opposed to volcanic) igneous rocks having the texture (but not necessarily the composition) of granite.

GYMNOSPERM: A class of fernlike or coniferlike plants whose seeds are not enclosed in an ovary.

HABITAT: A place in the environment where an organism lives.

HADROSAUR: A group of herbivorous dinosaurs commonly referred to as "duck-billed" dinosaurs.

HAEMAL ARCH: Arch of bone on the ventral side of some caudal vertebrae that protects blood vessels.

HISTOLOGICAL: Pertaining to tissue.

HOGBACK: In geology, a long linear ridge caused by folded or tilted strata resistant to erosion.

HUMERUS: The upper bone of the forelimb.

HYPSILOPHODONT: A fleet-footed, relatively small, herbivorous dinosaur with slender hind limbs and smaller front limbs.

ICE AGE: *See* Pleistocene.

ICHNOGENERA: Organisms known only from their tracks or trails.

ICHTHYOSAUR: One of a group of highly evolved, streamlined, fish-shaped, seagoing reptiles that have been found only in Mesozoic rocks.

INTEGUMENT: An outer covering, such as skin, hide, husk, rind, or shell.

ISCHIUM: A backward-projecting bone found on the ventral side of the hip girdle.

ISLAND ARC: A curving chain of islands containing volcanoes adjacent to a trench.

K-T BOUNDARY: The boundary between the Cretaceous (K) and Tertiary (T) Periods when, 65 million years ago, a vast majority of the species of life on Earth went extinct.

LENS: In geology, a lens-shaped (as in magnifying glass lens) bed of sediment.

LIMY: Having significant amounts of lime (calcium carbonate).

LITHIFICATION: The process of turning to stone.

LITHIFIED: Turned to stone.

MAGMA: Hot, liquid rock.

MANDIBLE: The lower jaw.

MANTLE: The area of rock rich in iron and magnesium, as well as some magma, beneath the crust of the earth and above the iron core.

MATRIX: In geology, the surrounding material in which a rock, mineral, or fossil is embedded.

MAXILLA: The part of the skull adjacent to the mandible, usually containing teeth; the upper jaw.

METACARPAL: Referring to the midbones of the hand or foot of the forelimb of vertebrates between the fingers (or toes) and the wrist (or ankle).

METAMORPHIC: Referring to rocks that have been formed by time, heat, pressure, and chemically active fluids.

METATARSAL: Referring to the midbones of the rear-limb foot that lie between the toes and the ankle.

MICROSCARP: A tiny clifflike structure.

MOLLUSK: One of a group (phylum) of invertebrates most of which usually have shells, including clams, mussels, oysters, pectins (scallops), snails, abalone, chambered nautilus, ammonites, and octopus.

MONKEY PUZZLE TREE: Common name for the primitive conifers in the family Araucariaceae (found today as natives only in the southern hemisphere). The genus *Araucaria* includes both the monkey puzzle tree and the Norfolk Island pine, which are common ornamental trees.

MONOPHYLETIC: Composed of a single taxon and all of its descendants.

MORPHOLOGY: The study of shapes; as in geomorphology (the shapes of the earth's surface) or the structure of fossil animal or plant remains.

MOSASAUR: A large seagoing lizard that lived only in the Cretaceous Period.

NARES: The nostrils of vertebrates.

NARIAL: Pertaining to the nostrils.

NICHE: An organism's role in its habitat.

ORNITHISCHIAN: One of two main categories of dinosaurs (the other being the saurischians) described mainly from the structure of the bones of the hip girdle. The members of this group had hips resembling birds ("bird-hipped"), with the pubis lying more or less parallel to the center of the pelvis, and all were herbivorous.

ORNITHOMIMID: A member of a group of very fast-running dinosaurs, known as the "ostrich dinosaurs"; they had a beaklike facial structure, big eyes, and a large brain enclosed in a small birdlike skull.

OSSIFIED: Converted to bone.

PALEOENVIRONMENT: Ancient environment.

PALEOFAUNA: Ancient animals.

PALEONTOLOGY: The study of prehistoric life through fossils.

PANGAEA: The supercontinent that existed during the early Mesozoic Era when all of the continents were a single landmass.

PECTORAL: Referring to the shoulder area.

PERIOTIC SINUS: A cavity situated in the vicinity of the ear.

PHALANX (PL. PHALANGES): Any of the bones that form the fingers or toes.

PLATE: An individual segment of the crust of the earth that may be bounded by such features as ridges, transform faults, or subduction zones.

PLATE TECTONICS: A theory that attempts to explain how various segments of Earth's crust move slowly about the globe and interact, producing seismic activity, volcanism, and mountain building.

PLEISTOCENE: An epoch of geologic time in the Cenozoic Era that began approximately 2.5 to 1.6 million years ago and ended about twenty thousand to ten thousand years ago. It is commonly known as the Ice Age.

PLESIOSAUR: A type of seagoing reptile that lived in the Jurassic and Cretaceous Periods. It had four paddle-shaped flippers, a long or medium-length neck, and a relatively small tooth-studded skull adapted to catch fish and other swimming food.

POSTCRANIAL: Referring to the body of an animal behind the skull.

POSTNARIAL: Referring to the area in back of the nostrils.

PREMAXILLA: The area in front of the cheek teeth of the lower jaw.

PRENATAL: Taking place or existing before birth.

PREPARATOR: In paleontology, a person who extracts fossils from their surrounding rock matrix, cleans them, then assembles the parts, if necessary.

PROXIMAL: Pertaining to the portion of something (such as the end of a bone) that is nearer to the body rather than that portion that is away from it (distal).

PTEROSAUR: One of a group of flying reptiles that lived during the Mesozoic Era.

PTERYGOID: A bone projecting from and lateral to the ventral part of the skull at the rear of the palate and behind the vomer.

PUBIS: Ventral, usually forward-projecting, portion of the hip girdle. This is the key bone separating ornithischian from saurischian dinosaurs.

PYROCLASTIC: Meaning "hot fragments," this word refers to rocks that have been ejected from volcanoes during eruptions.

RADIOLARIA: A one-celled, animal-like form of plankton that makes an ornate skeleton (test) of silicon dioxide. These tests are sometimes used for dating their enclosing rocks.

RADIOLARIAN CHERT: A form of chert (jasper) composed largely out of the siliceous skeletons of radiolaria.

REEFAL: Pertaining to reefs, as in a coral reef.

RIFT: A linear spreading center usually found along a midocean mountain chain.

RIPARIAN: Referring to the environment bordering a river, stream, or lake.

RIP-UP CLAST: A fragment of rock or sediment that is torn from the seafloor and carried along by an undersea landslide or turbidity current.

ROSTRUM: The beaklike projection on some fish, marine mammals, or reptiles (such as the sword of a swordfish or the snout of a porpoise).

SAURIAN: Having the characteristics of crocodiles, lizards, or dinosaurs. Used loosely to describe any large reptilian creature.

SAURISCHIAN: One of the two broadly based groups of dinosaurs (the other being the ornithischians) described mainly from the structure of the bones of the hip girdle ("lizard-hipped"). In saurischian dinosaurs, which include the large sauropods and all of the theropods, the pubis slants forward and down from the center of the pelvis.

SAUROPOD: A member of the group of quadrupedal, long-necked, long-tailed, plant-eating saurischian dinosaurs belonging to the suborder Sauropodomorpha.

SCAPULA: The bladelike bone of the shoulder, commonly known as the shoulder blade, that articulates with the humerus. In reptiles the scapula is often fused to the coracoid bone to form the scapulocoracoid.

SCUTE: A usually small, protective, platelike bone embedded in the skin.

SEDIMENTARY: Referring to a major group of rocks made out of clastic and/or chemical debris that is pressed and/or cemented together.

SHALE: A sedimentary rock composed of clastic debris that is smaller than sand, such as mud, silt, or clay.

SILICEOUS: Referring to a substance, such as a rock, that contains abundant silica, usually in the form of silicon dioxide (quartz).

SILICIFIED: Referring to something that has been converted to, or impregnated by, silicon dioxide.

STRATIGRAPHY: The study of the origin, composition, distribution, and succession of rock strata.

STRATUM (PL. STRATA): A layer of rock.

STRIKE VALLEY: A valley formed by the erosion of softer layers within folded or tilted rock strata, which hence parallels long linear ridges (hogbacks) made out of the more resistant strata.

SUBDUCTION: The tectonic process by which one crustal plate descends beneath another.

SUBMARINE FAN: A sloping, fan-shaped depositional area found usually at the mouth of a submarine canyon.

TAXON: A named group of organisms.

TAXONOMY: The practice and theory of classifying life forms into groups containing like characteristics.

TERRESTRIAL: Pertaining to areas of the earth that are above sea level.

THALATTOSAUR: One of a group of marine reptiles found only in Triassic rocks that are similar to primitive ichthyosaurs and not too unlike crocodiles in form. They had long and flattened eel-like tails and limbs that were most likely webbed and paddlelike, with grasping claws.

THEROPOD: A meat-eating dinosaur, partly characterized by a lower jaw that contained an extra joint. They had bladelike serrated teeth, were bipedal, and the outer fingers on their hands often were reduced in size or number.

TIBIA: One of the two bones, usually the larger, of the hind leg below the knee.

TRENCH: In plate tectonics, a linear deep area of the ocean floor, usually on the ocean side of volcanic mountains.

TURBIDITE: A sedimentary deposit, usually graded from coarse on the bottom to fine at the top, formed by deposition left from a turbidity current.

TURBIDITY CURRENT: A fast-flowing, dense, bottom-hugging current formed from a combination of sediment and water that usually flows from a nearshore environment into deeper water.

TYPE SPECIMEN: The single specimen of a life form or fossil on which the description of the species is based, serving as a permanent reference for the application of the name.

ULNA: One of the two long bones, usually the larger, found below the knee or elbow of the front limb.

VARANID: Also called platynotans, the varanids are the most advanced of all lizards in achieving an active, predaceous way of life and in terms of their large size. They have a hinged jaw with replacement teeth that come in between the mature teeth, and a tongue with bifid (cleft) sensory tips. The Komodo dragon is a varanid.

VASCULARIZED: Containing many blood vessels.

VERTEBRATE: Subphylum of the phylum Chordata. Vertebrates usually have bone and/or cartilage as supporting structure and usually have vertebrae and a skull.

VOMER: One of the bones of the palate.

MUSEUMS AND WEBSITES TO VISIT TO LEARN MORE ABOUT CALIFORNIA MESOZOIC REPTILE FOSSILS

California Academy of Sciences, in Golden Gate Park
   San Francisco, California
   www.calacademy.org/
   Numerous fossils

Life Science Building
   Monterey Peninsula College, Monterey, California
   Mosasaur

McLane Hall at California State University, Fresno
   Fresno, California
   Cast of *Hydrotherosaurus alexandrae*

Natural History Museum of Los Angeles County
   900 Exposition Blvd., Los Angeles, California 90007
   213-763-DINO
   www.nhm.org
   Numerous fossils

Orange County Natural History Museum, at Ralph B. Clark Park
   Buena Vista, California
   www.ocnha.mus.ca.us/
   Fossils

San Bernardino County Museum of Natural History
   San Bernardino, California
   www.co.san-bernardino.ca.us/museum/
   Fossils and trackways

San Diego Museum of Natural History
   San Diego, California
   www.sdnhm.org/
   Numerous fossils

Sierra College Natural History Museum, in Sewell Hall
Rocklin, California
www.sierramuseum.org
Numerous fossils

University of California Museum of Paleontology, in the Valley Life Science Building
Berkeley, California
www.ucmp.berkeley.edu/
Numerous fossils

W. M. Keck Museum, in the Mackay School of Mines Building, University of Nevada
Reno, Nevada
www.mines.unr.edu/museum/
Cast of *Hydrotherosaurus alexandrae*

BIBLIOGRAPHY

ARCHIVAL MATERIAL

Alexander, A. M. Misc. letters and papers. Bancroft Library, University of California, Berkeley.

Merriam, J. C. Misc. letters and papers. Bancroft Library, University of California, Berkeley.

————. Misc. letters. California Institute of Technology Archives, Pasadena.

Stock, C. Misc. letters. California Institute of Technology Archives, Pasadena.

————. Misc. letters and papers. Chester Stock Library of the Page Museum, Hancock Park, Los Angeles.

Welles, S. P. Unpub. field notes. Museum of Paleontology, University of California, Berkeley.

PUBLICATIONS

Andrews, C. W. 1910–1913. *Descriptive Catalogue of the Marine Reptiles of the Oxford Clay: Based on the Leeds Collection in the British Museum.* 2 vols. London: British Museum (Natural History), Department of Geology.

Anonymous. 1902. A monster in the rocks; strange discovery by scientists up at Clipper Gap. *Sunday News* (Sacramento), January 19, 1902, 8:1.

Bardet, N. 1992. Stratigraphic evidence for the extinction of the ichthyosaurs. *Terra Nova* 4: 649–656.

Barsbold, R. 1997. Mongolian dinosaurs. In *The Encyclopedia of Dinosaurs,* ed. P. J. Currie and K. Padian, 447–450. New York: Academic Press.

Barsbold, R., and H. Osmolska. 1990. Ornithomimosauria. In *The Dinosauria,* ed. D. B. Weishampel, P. Dodson, and H. Osmolska, 225–244. Berkeley: University of California Press.

Bell, C. M., and K. Padian. 1995. Pterosaur fossils from the Cretaceous of Chile: Evidence for a pterosaur colony on an inland desert plain. *Geol. Mag.* 132(1): 31–38.

Bell, G. L. 1993. A phylogenetic revision of Mosasauroidea (Squamata). Ph.D. diss., Univ. Texas Austin.

————. 1996. Decapitating a mosasaur. *Discover* 17(2).

————. 1997. A phylogenetic revision of North American and Adriatic Mosasauroidea. In *Ancient Marine Reptiles,* ed. J. M. Callaway and E. L. Nicholls, 293–332. New York: Academic Press.

Bell, G. L., and A. L. Sheldon. 1996. The first direct evidence of live birth in Mosasauridae (Squamata): Exceptional preservation in the Pierre Shale of South Dakota (abstr.). *J. Vert. Paleontol.* 16(3): 21A.

Benton, M. J. 1998. *The Reign of Reptiles.* London: Quantum Books.

Blackwelder, E. 1965. James Perrin Smith, Nov. 27, 1864–Jan. 1, 1931. *Natl. Acad. Sci., Biog. Memoirs* 38: 295–308.

Bogen, N. L. 1984. Stratigraphic and sedimentologic evidence of a submarine island-arc volcano in the lower Mesozoic Penon Blanco and Jasper Point Formations, Mariposa County, California. *Geol. Soc. Am. Bull.* 95: 1322–1331.

Boulenger, G. A. 1904. A remarkable ichthyosaurian anterior paddle. *Proc. Zool. Soc. of London* 1: 424–426.

Brett-Surman, M. K. 1997. Ornithopods. In *The Complete Dinosaur,* ed. J. O. Farlow and M. K. Brett-Surman, 330–346. Bloomington: Indiana University Press.

Brewer, W. H. 1966. *Up and Down California in 1860–1864: The Journal of William H. Brewer.* Ed. F. P. Farquhar. Berkeley: University of California Press.

Brinkman, D. B. 1998. The skull and neck of the Cretaceous turtle *Basilemys* (Trionychoidea, Nanhsiungchelyidae), and the interrelationships of the genus. *Paludicola* 1(4): 150–157.

Brochu, C. A. 1999. Phylogenetics, taxonomy, and historical biogeography of Alligatoroidea. In *Cranial morphology of* Alligator mississippiensis *and phylogeny of Alligatoroidea,* by T. Rowe, A. B. Christopher, and K. Kyoko. *J. Vert. Paleontol.,* supp. to 19(2): 9–100.

Brodkorb, P. 1976. Discovery of a Cretaceous bird, apparently ancestral to orders Coraciiformes and Piciformes (Aves: Carinatae). *Smithsonian Contributions to Paleobiol.,* no. 27: 67–73.

Brown, B. 1904. Stomach stones and food of Plesiosaurs. *Science,* n.s. 20(501): 184–185.

Brown, D. S. 1981. The English upper Jurassic Plesiosauroidea (Reptilia) and a review of the phylogeny and classification of the Plesiosauria. *Bull. Br. Mus. Nat. Hist. (Geol.)* 35(4): 253–347.

Buchanan, R. C., ed. 1984. *Kansas Geology: An Introduction to Landscapes, Rocks, Minerals, and Fossils.* Lawrence: University Press of Kansas.

Buffetaut, E. 1979. Jurassic marine crocodilians (Mesosuchia: Teleosauridae) from central Oregon: First record in North America. *J. Paleontol.* 53: 210–215.

Buwalda, J. P. 1951. Chester Stock (1892–1950). *Bull. Am. Assoc. Petrol. Geol.* 35(3): 775–778.

Caldwell, M. W. 1997. Limb osteology and ossification patterns in *Cryptoclidus* (Reptilia: Plesiosauroidea), with a review of Sauropterygian limbs. *J. Vert. Paleontol.* 17(2): 295–297.

Callaway, J. M. 1986. Systematic revision of the Shastasauridae (Reptilia, Ichthyosauria) with implications for the origins of Ichthyosaurs. *N. Am. Paleontol. Conv. IV, Abstracts with Programs,* A8.

———. 1987. Ancestry and phylogeny of ichthyosaurs (abstr.). *J. Vert. Paleontol.* 7(3): 13A.

Callaway, J. M., and J. A. Massare. 1989a. Geographic and stratigraphic distribution of the Triassic Ichthyosauria (Reptilia; Diapsida). *Neues Jahrbuch für Geologie und Paläontologie, Abhandlungen* 178: 37–58.

———. 1989b. *Shastasaurus altispinus* (Ichthyosauria, Shastasauridae) from the upper Triassic of the El Antimonio district, northwestern Sonora, Mexico. *J. Vert. Paleontol.* 63(6): 930–939.

Callaway, J. M., and E. L. Nicholls. 1997. *Ancient Marine Reptiles.* New York: Academic Press.

Camp, C. L. 1942a. *California Mosasaurs.* Mem. Univ. Calif. 13. Berkeley and Los Angeles: University of California Press.

———. 1942b. Ichthyosaur rostra from central California. *J. Paleontol.* 16(3): 362–371.

———. 1951. *Plotosaurus,* a new generic name for *Kolposaurus* Camp, preoccupied. *J. Paleontol.* 25: 822.

———. 1976. Vorläufige Mitteilung über große Ichthyosaurier aus der oberen Trias von Nevada. *Österische Akademie der Wissenschaften, Mathematisch-naturwissenschaftliche Klasse, Sitzungsberichte,* Abt. 1, 185: 125–134.

———. 1980. Large ichthyosaurs from the upper Triassic of Nevada. *Palaeontographica* A170: 139–200.

Campbell, C., and T. Lee. 2001. "Tails of Hoffmanni": Mosasaur fossils in tsunami deposit at K/T boundary of southeast Missouri (abstr.). *J. Vert. Paleontol.* 21(3): 37A.

Carpenter, K. 1982. Skeletal and dermal armor reconstruction of *Euoplocephalus tutus* (Ornithischia: Ankylosauridae) from the Late Cretaceous Oldman Formation of Alberta. *Can. J. Sci.* 19: 689–697.

———. 1989. Dolichorhynchops / = Trinacromerum. *J. Vert. Paleontol.* 9, supp. 3: 15A.

———. 1997a. Ankylosauria. In *The Encyclopedia of Dinosaurs,* ed. P. J. Currie and K. Padian, 16–19. New York: Academic Press.

———. 1997b. Tyrannosauridae. In *The Encyclopedia of Dinosaurs,* ed. P. J. Currie and K. Padian, 766–768. New York: Academic Press.

Carpenter, K., and D. Lindsey. 1980. The dentary of *Brachychampsa montana* Gilmore (Alligatorinae: Crocodylidae), a Late Cretaceous turtle-eating alligator. *J. Paleontol.* 54: 1213–1217.

Carroll, R. L. 1988. *Vertebrate Paleontology and Evolution.* New York: W. H. Freeman.

Case, G. R. 1982. *A Pictorial Guide to Fossils.* New York: Van Nostrand Reinhold.

Chatterjee, S. 2001. Flight of pterosaurs (abstr.). *J. Vert. Paleontol.* 21(3): 40A.

Chiappe, L. 1997. Aves. In *The Encyclopedia of Dinosaurs,* ed. P. J. Currie and K. Padian, 32–38. New York: Academic Press.

Christe, G., and R. P. Hilton. 2001. Vertebrate fossils from the upper? Jurassic Trail Formation, Kettle Rock Sequence: First recorded evidence of dinosaur remains in the Sierra Nevada, Calif. *Geol. Soc. Am., Abstracts with Programs* 33(3): A-53.

Clemens, W. A., and L. G. Nelms. 1993. Paleoecological implications of Alaskan terrestrial vertebrate fauna in latest Cretaceous time at high paleolatitudes. *Geology* 21: 503–506.

Coombs, W. P., Jr. 1972. The bony eyelid of *Euoplocephalus* (Reptilia, Ornithischia). *J. Paleontol.* 46: 637–651.

Coombs, W. P., Jr., and T. A. Deméré. 1996. A Late Cretaceous nodosaurid ankylosaur (Dinosauria: Ornithischia) from marine sediments of coastal California. *J. Paleontol.* 70: 311–326.

Currie, P. J. 1985. Cranial anatomy of *Stenonychosaurus inequalis* (Saurischia, Theropoda) and its bearing on the origin of birds. *Can. J. Earth Sci.* 22: 247–265.

———. 1997a. Dromaeosauridae. In *The Encyclopedia of Dinosaurs,* ed. P. J. Currie and K. Padian, 194–195. New York: Academic Press.

———. 1997b. Theropoda. In *The Encyclopedia of Dinosaurs,* ed. P. J. Currie and K. Padian, 731–737. New York: Academic Press.

———. 1997c. Theropods. In *The Complete Dinosaur,* ed. J. O. Farlow and M. K. Brett-Surman, 216–233. Bloomington: Indiana University Press.

Currie, P. J., and P. Dodson. 1984. Mass death of a herd of ceratopsian dinosaurs. In *3d Symp. Mesozoic Terrest. Ecosyst.,* ed. W.-E. Reif and F. Westphal, 61–66. Tübingen, Ger.: Attempto.

Currie, P. J., and K. Padian, eds. 1997. *The Encyclopedia of Dinosaurs.* New York: Academic Press.

Czerkas, S. J., and S. A. Czerkas. 1991. *Dinosaurs: A Global View.* New York: Mallard Press.

Darby, D. G., and R. W. Ojakangas. 1980. Gastroliths from an upper Cretaceous plesiosaur. *J. Paleontol.* 54: 548.

DeCourten, F. L. 1997. Dinosaurs in California? *Pacific Discovery* 50: 26–31.

Deméré, T. A. 1985. Dinosaurs of California. *Environment Southwest* (San Diego Nat. Hist. Mus.), no. 509: 15–17.

———. 1988. An armored dinosaur from Carlsbad. *Environment Southwest* (San Diego Nat. Hist. Mus.), no. 523: 12–15.

Dickinson, W. R. 1976. Sedimentary basins developed during evolution of Mesozoic-Cenozoic arc-trench system in western North America. *Can. J. Earth Sci.* 13: 1268–1287.

Dodson, P. 1975. Taxonomic implications of relative growth in lambeosaurine hadrosaurs. *Syst. Zool.* 24: 37–54.

———. 1997a. Ceratopsia. In *The Encyclopedia of Dinosaurs,* ed. P. J. Currie and K. Padian, 106. New York: Academic Press.

———. 1997b. Distribution and diversity. In *The Encyclopedia of Dinosaurs,* ed. P. J. Currie and K. Padian, 186–188. New York: Academic Press.

———. 1997c. Neoceratopsia. In *The Encyclopedia of Dinosaurs,* ed. P. J. Currie and K. Padian, 473–478. New York: Academic Press.

———. 1997d. Paleoecology. In *The Encyclopedia of Dinosaurs,* ed. P. J. Currie and K. Padian, 515–519. New York: Academic Press.

Downs, T. 1968. *Fossil Vertebrates of Southern California.* Berkeley: University of California Press, 61.

Drewry, G. E., A. T. S. Ramsey, and G. Smith. 1974. Climatically controlled sediments, the geomagnetic field, and tradewind belts in Phanerozoic time. *J. Geol.* 82(5): 531–553.

Dupras, D. 1988. Ichthyosaurs of California, Nevada, and Oregon. *Calif. Geol.* 41: 99–107.

Eaton, C. F. 1910. Osteology of pteranodon. *Mem. Conn. Acad. Arts and Sci.* 2: 1–38.

Evans, J. R. 1958. Geology of the Mescal Range, San Bernardino County, California. Master's thesis, Univ. So. Calif.

———. 1971. *Geology and mineral deposits of the Mescal Range Quadrangle, San Bernardino County, California.* Calif. Div. Mines and Geol. Map, Sheet 17.

Farlow, J. O., and M. K. Brett-Surman. *The Complete Dinosaur.* Bloomington: Indiana University Press.

Fiero, B. 1986. *Geology of the Great Basin.* Reno: University of Nevada Press.

Fleck, R. J., and R. E. Reynolds. 1996. Mesozoic stratigraphic units of the eastern Mescal Range, southeastern California. *San Bernardino Co. Mus. Assn. Quarterly* 43(1,2): 49–54.

Ford, T. L., and Kirkland, J. I. 2001. Carlsbad ankylosaur (Ornithischia, Ankylosauria): An ankylosaurid and not a nodosaurid. In *The Armored Dinosaurs,* ed. K. Carpenter. Bloomington: Indiana University Press.

Forster, C. A. 1997. Hadrosauridae. In *The Encyclopedia of Dinosaurs,* ed. P. J. Currie and K. Padian, 293–298. New York: Academic Press.

Forster, C. A., S. D. Sampson, L. M. Chiappe, and D. W. Krause. 1998. The theropod ancestry of birds: New evidence from the Late Cretaceous of Madagascar. *Science* 279: 1915–1918.

Forster, C. A., and P. C. Sereno. 1997. Marginocephalians. In *The Complete Dinosaur,* ed. J. O. Farlow and M. K. Brett-Surman, 317–329. Bloomington: Indiana University Press.

Foster, D. E. 1980. *Osteopigis* sp., a marine turtle from the Late Cretaceous Moreno Formation of California. *PaleoBios* (Mus. Paleontol., Univ. Calif., Berkeley), no. 34 (Dec. 18): 1–14.

Galton, P. M. 1974. The ornithischian dinosaur Hypsilophodon from the Wealden of the Isle of Wight. *Bull. Br. Mus. Nat. Hist. (Geol.)* 25: 1–152.

Gilmore, C. W. 1928. A new pterosaurian reptile from the marine Cretaceous of Oregon. *Proc. U.S. Natl. Mus.* 73(24): 1–5.

Grady, W. 2001. Underwater dinosaurs: Elizabeth Nicholls quarries the world's largest ichthyosaur. *Can. Geog.,* March/April, 72–80.

Gregory, J. T. 1995. Annie Montague Alexander. *Newsletter, Univ. Calif. Mus. Paleontol.,* Oct., 2–3.

———. 1997. Paleontology Loses Sam Welles. *Newsletter, Univ. Calif. Mus. Paleontol.,* Sept., 3.

———. 1998. Samuel P. Welles obituary. *Soc. Vert. Paleontol.,* no. 173: 77–79.

Grinnell, H. W. 1958. *Annie Montague Alexander.* Berkeley: Grinnell Naturalists Soc., Univ. Calif. Mus. Vert. Zool.

Hannah, J. L., and E. M. Moores. 1986. Age relationships and depositional environments of Paleozoic strata, northern Sierra Nevada, California. *Geol. Soc. Am. Bull.* 97: 787–797.

Hernandez-Rivera, R. 1997. Mexican dinosaurs. In *The Encyclopedia of Dinosaurs,* ed. P. J. Currie and K. Padian, 433–437. New York: Academic Press.

Hesse, C. J., and S. P. Welles. 1936. The first record of a dinosaur from the West Coast. *Science* 84: 157.

Hilton, R. P. 1975. The geology of the Ingot-Round Mountain area, Shasta County, California. Master's thesis, Calif. State Univ., Chico.

Hilton, R. P., and P. J. Antuzzi. 1997. Chico Formation yields clues to Late Cretaceous paleoenvironments in California. *Calif. Geol.* 50: 135–144.

Hilton, R. P., F. L. DeCourten, and P. G. Embree. 1995. First California dinosaur north of Sacramento. *Calif. Geol.* 48: 99–102.

Hilton, R. P., F. L. DeCourten, M. A. Murphy, P. U. Rodda, and P. G. Embree. 1997. An early Cretaceous ornithopod dinosaur from California. *J. Vert. Paleontol.* 17(3): 557–560.

Hilton, R. P., E. S. Göhre, P. G. Embree, and T. A. Stidham. 1999. California's first fossil evidence of Cretaceous winged vertebrates. *Geol.,* July-Aug., 4–10.

Hirayama, R. 1997. Distribution and diversity of Cretaceous chelonioids. In *Ancient Marine Reptiles,* ed. J. M. Callaway and E. L. Nicholls, 225–241. New York: Academic Press.

Holden, C. 2002. Fossil vomit. *Science* 295: 1459.

Horner, J. R. 1979. Upper Cretaceous dinosaurs from the Bearpaw Shale (marine) of south-central Montana with a checklist of Upper Cretaceous remains from marine sediments in North America. *J. Paleontol.* 53: 566–578.

———. 1997a. Behavior. In *The Encyclopedia of Dinosaurs,* ed. P. J. Currie and K. Padian, 45–48. New York: Academic Press.

———. 1997b. Willow Creek Anticline. In *The Encyclopedia of Dinosaurs,* ed. P. J. Currie and K. Padian, 786. New York: Academic Press.

Horner, J. R., and R. Makela. 1979. Nests of juveniles provide evidence of family structure among dinosaurs. *Nature* 282: 296–298.

Howell, D. G., and K. McDougall, eds. 1978. *Mesozoic Paleogeography of the Western United States.* Pacific Coast Paleogeography Symp. 2. Los Angeles: Pacific Sec., Soc. Econ. Paleontol. Mineral.

Hua, S., and E. Buffetaut. 1997. Crocodylia. In *Ancient Marine Reptiles,* ed. J. M. Callaway and E. L. Nicholls, 357–374. New York: Academic Press.

Hutchinson, J. R., and K. Padian. 1997. *Coelurosauria.* In *The Encyclopedia of Dinosaurs,* ed. P. J. Currie and K. Padian, 45–48. New York: Academic Press.

Jones, K. 1902. Notes of Miss Katherine Jones, Paleontological Expedition, University of California, Shasta County, California, June 16 to July 13, 1902. Univ. Calif., Berkeley, Mus. Vert. Zool. Archives.

Kase, T., P. Johnson, A. Seilacher, and J. Boyce. 1998. Alleged mosasaur bite marks on Late Cretaceous ammonites are limpet (patellogastropod) home scars. *Geology* 26: 947–950.

Kilmer, F. H. 1963. Cretaceous and Cenozoic Stratigraphy and Paleontology, El Rosario Area, Baja California, Mexico. Ph.D. diss., Univ. Calif., Berkeley.

Kooser, M. [1985?] Paleocene Plesiosaur? In *Geologic Investigations along Interstate 15, Cajon Pass to Manix Lake, California,* ed. R. E. Reynolds, 43–48. Redlands: San Bernardino County Museum.

Kosch, B. F. 1990. A revision of the skeletal reconstruction of *Shonisaurus populanis* (Reptilia: Ichthyosauria). *J. Vert. Paleontol.* 10: 512–514.

Krutch, J. W. 1961. *The Forgotten Peninsula: A Naturalist in Baja.* New York: William Morrow.

Kuhn, O. 1934. Ichthyosauria. In *Fossilium Catalogus.* Vol. 1: *Animalia,* ed. W. Quenstedt. Pars 63. Berlin: Junk.

Kuhn-Schnyder, E. 1974. Die Triasfauna der Tessiner Kalkalpen. *Neujahrsblatt, Naturf. Ges. Zürich* 176: 1–119.

Kurtz, W. 2001. Trackway evidence for gregarious ankylosaurs from the Dakota Group (Cretaceous, Albian) of Colorado (abstr.). *J. Vert. Paleontol.* 21(3): 70A.

Lambert, D. 1993. *The Ultimate Dinosaur Book.* London: Dorling Kindersley.

Lander, E. B. Eastern Transportation Corridor paleontologic resource impact mitigation program—Final technical report findings. Paleo Environmental Associates, Inc., Projects 94-23 and 99-11. Prepared for Foothill/Eastern Transportation Corridor Agency and Raytheon Infrastructure Services, Inc.

Langston, W., Jr. 1956. The shell of *Basilemys variolosa* (Cope). *Bull. Natl. Mus. of Can.* 147: 155–163.

Langston, W., Jr., and M. H. Oakes. 1954a. Hadrosaurs in Baja California. *Abstr. Geol. Soc. Am. Bull.* 65(12), pt. 2: 1344.

———. 1954b. Hadrosaurs in Baja California. *Cord. Sec. Abstr. Geol. Soc. Am. Bull. Official Program, Seattle, March 26–27, 1954.*

Leidy, J. 1856. Notice of remains of extinct reptiles and fishes, discovered by Dr. F. V. Hayden in the badlands of the Judith River, Nebraska Territory. *Proc. Acad. Nat. Sci., Philadelphia* 8: 72–73.

Lessem, D., and D. F. Glut. 1993. *The Dinosaur Society's Dinosaur Encyclopedia.* New York: Random House.

Lillegraven, J. A. 1970. El Gallo Project, pertinent notes: Summer 1970. Los Angeles Co. Mus. Nat. Hist. Typescript.

Longrich, N. R. 2001. Secondarily flightless Maniraptorian Theropods? (abstr.). *J. Vert. Paleontol.* 21(3): 74A.

Lucas, L. G., and R. E. Reynolds. 1991. Late Cretaceous(?) plesiosaurs from Cajon Pass, California. *San Bernardino Co. Mus. Assn. Quarterly* 38(3): 52–53.

———. 1993. Putative Paleocene plesiosaurs from Cajon Pass, California, U.S.A. *Cret. Res.* 14: 107–111.

Macdonald, D. J. 1967. Mission to Mexico. Los Angeles Co. Mus. Nat. Hist. Typescript.

Marsh, O. C. 1880. *Odontornithes: A Monograph on the Extinct Toothed Birds of North America. United States Geological Exploration of the Fortieth Parallel. Clarence King, Geologist-in-Charge.* Washington, D.C.: U.S. Govt. Printing Office.

Massare, J. A., and J. M. Callaway. 1987. The ecology of Triassic ichthyosaurs (abstr.). *J. Vert. Paleontol.* 7(3): 20A.

Matthew, W. D. 1929. John Campbell Merriam, new president of the Carnegie Institute. *Nat. Hist.* 20: 253–254.

McAuliffe, K. 1995. Elephant seals, the champion divers of the deep. *Smithsonian,* Sept., 45–56.

McGowan, C. 1976. The description and phenetic relationships of a new ichthyosaur genus from the Upper Jurassic of England. *Can. J. Earth Sci.* 13: 668–683.

———. 1983a. *Dinosaurs, Spitfires, and Sea Dragons.* Cambridge, Mass.: Harvard University Press.

———. 1983b. *The Successful Dragons, a Natural History of Extinct Reptiles.* Sarasota, Fla.: Samuel Stevens and Co.

McMath, V. E. 1966. Geology of the Taylorsville Area, northern Sierra Nevada. In *Geology of Northern California*, ed. E. H. Bailey, 173–183. Calif. Div. Mines and Geol., Bull. 190.

McNassor, C. 1992. Chester Stock, 1892–1950, a memorable fossil hunter. *Terra* 31(1): 28–34.

Merriam, J. C. 1895. On some reptilian remains from the Triassic of northern California. *Am. J. Sci.* 1: 55.

———. 1902a. Triassic Ichthyopterygia from California and Nevada. *Univ. Calif. Publications, Dept. Geol. Bull.* 3(4): 63–108.

———. 1902b. Triassic Reptilia from northern California. *Science*, n.s., 15: 411.

———. 1903a. New Ichthyosauria from the Upper Triassic of California. *Univ. Calif. Publications, Dept. Geol. Bull.* 3(12): 249–263.

———. 1903b. Recent literature on Triassic Ichthyosauria. *Science*, n.s., 18: 311–312.

———. 1904. A new marine reptile from the Triassic of California. *Univ. Calif. Publications, Dept. Geol. Bull.* 3: 419–421.

———. 1905a. A new group of marine reptiles from the Triassic of California. *Comptes rendus du 6 congrès internationale de zoologie (Berne, 1904)*, 247–248.

———. 1905b. The thalattosauria: A group of marine reptiles from the Triassic of California. *Calif. Acad. Sci. Mem.* 5: 1–52.

———. 1905c. The types of limb-structure in Triassic Ichthyosauria. *Am. J. Sci.*, 4th ser., 19: 23–30.

———. 1908a. Notes on the osteology of the thalattosaurian genus *Nectosaurus*. *Univ. Calif. Publications, Dept. Geol. Bull.* 5(13): 217–223.

———. 1908b. *Triassic Ichthyosauria, with Special Reference to the American Forms.* Mem. Univ. Calif. 1, no. 1. Berkeley: University of California Press.

———. 1910. The skull and dentition of a primitive ichthyosaurian from the Middle Triassic. *Univ. Calif. Publications, Dept. Biol. Bull.* 5(25): 381–390.

———. 1911. An account of the past ten years' work on the history of marine reptiles. *Paleontol. Soc. Proc.*, 221–223. (Same as *GSA Bull.* 1912.)

———. 1912. Marine reptiles: Ten years' progress in vertebrate paleontology. *Geol. Soc. Am. Bull.* 23: 221–223.

Merriam, J. C., and B. Clark. [1911]. The skull of *Thalattosaurus*. (Unpub. [lost?]; mentioned in account by Merriam to A. Alexander, Jan. 13.)

Merriam, J. C., and C. W. Gilmore. 1928. An ichthyosaurian reptile from the marine Cretaceous of Oregon. In *Contributions to Paleontology*, 3–4. Carnegie Inst. Wash. Pub. 393.

Merriam, J. C., et al. 1934. Continuation of paleontological researches. *Carnegie Inst. Wash. Yearbook*, no. 33 (Dec. 14), 302–313.

Mlynarski, M. 1972. *Zangerlia testudinimorpha* n. gen., n. sp.: A primitive land tortoise from the Upper Cretaceous of Mongolia. *Palaeontol. Polonica* 27: 85–92.

Molnar, R. E. 1974. A distinctive theropod dinosaur from the Upper Cretaceous of Baja California (Mexico). *J. Paleontol.* 48: 1009–1017.

Morris, W. J. 1965 (1971). Mesozoic and Tertiary vertebrates in Baja California. *Natl. Geog. Soc. Res. Rept.* 1965: 195–198.

———. 1966a. Investigations of Early Tertiary and Late Cretaceous mammals in Baja California, Mexico. Field Rept., summer 1966. Los Angeles Co. Mus. Nat. Hist. Typescript.

———. 1966b (1973). Mesozoic and Tertiary vertebrates in Baja California. *Natl. Geog. Soc. Res. Rept.* 1966: 197–209.

———. 1967a. Investigations of Mesozoic and Tertiary fossil vertebrates in Baja California. Field Rept., summer 1967. Los Angeles Co. Mus. Nat. Hist. Typescript.

———. 1967b. Late Cretaceous vertebrates from Baja California. *GSA Special Papers, Abstracts for 1967,* 341.

———. 1967c. Baja California: Late Cretaceous dinosaurs. *Science* 155: 1539–1541.

———. 1968a. Late Cretaceous dinosaurs from Baja California. *Abstr. Ann. Mtg., Geol. Soc. Am. Symp.: Middle America Paleobiol.*

———. 1968b. Late Cretaceous vertebrates from Baja California. *Abstr. Geol. Soc. Am.,* 341.

———. 1968c (1976). Mesozoic and Tertiary vertebrates of Baja California, 1968–1971. *Natl. Geog. Soc. Res. Rept.,* 305–316.

———. 1969a. National Geographic report for 1969: Investigations of Mesozoic and Tertiary fossil vertebrates in Baja California. Los Angeles Co. Mus. Nat. Hist. Typescript.

———. 1969b. A dinosaur vertebra from La Jolla, California. Oral report.

———. 1969c. A new lineage of Hadrosaurian dinosaurs. Oral report.

———. 1970. Late Cretaceous dinosaurs from Baja California. In *Pacific slope geology of northern Baja California and adjacent Alta California,* 67. Tulsa: Am. Assoc. of Petrol. Geol., Soc. Econ. Paleontol. Mineral., and Soc. Econ. Geophys.

———. 1971. National Geographic report for 1970: Investigations of Mesozoic and Tertiary fossil vertebrates in Baja California. Los Angeles Co. Mus. Nat. Hist. Typescript.

———. 1972. A giant hadrosaur from Baja California. *J. Paleontol.* 46: 777–779.

———. 1973a. A review of Pacific Coast hadrosaurs. *J. Paleontol.* 47: 551–561.

———. 1973b. Terrestrial avian fossils from Mesozoic strata, Baja California. *Natl. Geog. Soc. Res. Rept.* 14: 487–489.

———. 1974a. A paleontologic reconnaissance of Baja California, Mexico, 1974. *Natl. Geog. Soc. Res. Rept.* 15: 157–175.

———. 1974b. Upper Cretaceous "El Gallo" Formation and its vertebrate fauna. In *The Geology of Peninsular California,* ed. G. Gastil and J. Lillegraven. 49th Annual Meeting, Pacific Section AAPG-SEPM-SEG, April 24–28, 1974.

———. 1974c. National Geographic Society report for the project: Survey for terrestrial avian fossils, Upper Cretaceous, Baja California. Submitted January 1974 and covering field work from August to September 1973. Los Angeles Co. Mus. Nat. Hist. Typescript.

———. 1981. A new species of hadrosaurian dinosaur from the Upper Cretaceous of Baja California: *?Lambeosaurus laticaudus. J. Paleontol.* 55: 453–462.

———. 1982. California dinosaurs. In *Late Cretaceous depositional environments and paleogeography, Santa Ana Mountains, Southern California,* ed. D. J. Bottjer, I. P. Colburn, and J. D. Cooper. Los Angeles: Pacific Sec., Soc. Econ. Paleontol. Mineral.

Motani, R. 1999. Phylogeny of the Ichthyopterygia. *J. Vert. Paleontol.* 19: 472–495.

———. 2000. Rulers of the Jurassic seas. *Sci. Am.* 283: 52–59.

Motani, R., M. Nachio, and A. Tatsuro. 1998. Ichthyosaurian relationships illuminated by new primitive skeletons from Japan. *Nature* 393: 255–257.

Murphy, M. A., P. U. Rodda, and D. M. Morton. 1969. Geology of the Ono Quadrangle, Shasta and Tehama Counties, California. *Calif. Div. Mines and Geol. Bull.* 192: 1–28.

Nessov, L. A. 1984. Data on the Late Mesozoic turtles from the USSR. *Studia Geologie Salamanticensia.* Vol. Esp. 1: *Studia Palaecheloniologica* 1: 215–233.

Nessov, L. A., and L. B. Golovneva. 1990. [History of the flora, vertebrates, and climate in the late Senonian of the north-eastern Koriack.] In: *Continental Cretaceous of the USSR,* 191–212. Collection of Research Papers, Project 245, IUGS, Vladivostok. (In Russian).

Nicholls, E. L. 1994. The type materials of *Thalattosaurus* and *Nectosaurus* and the interrelationships of *Thalattosaurus. J. Vert. Paleontol.,* supp. to 14(3): 40.

———. 1999. A reexamination of *Thalattosaurus* and *Nectosaurus* and the relationships of the Thalattosaria (Reptilia: Diapsida). *PaleoBios* 19(1): 1–29.

———. 2000. A new genus of ichthyosaur from the Late Triassic of British Columbia and the problem with *Shastasaurus. Abstr. J. Vert. Paleontol.* 20(3): 60A.

Nicholls, E. L., and A. P. Russell. 1985. Structure and function of the pectoral girdle and forelimb of *Struthiomimus altus* (Theropoda: Ornithomimidae). *Paleontol.* 28: 643–677.

Nilsen, T. H. 1986. Cretaceous paleogeography of western North America. In *Cretaceous Stratigraphy of Western North America,* ed. P. L. Abbott, 1–40. Los Angeles: Pacific Sec., Soc. Econ. Paleontol. Mineral.

Norell, M. A., E. S. Gaffney, and L. Dingus. 1995. *Discovering Dinosaurs in the American Museum of Natural History.* Toronto: Knopf/Random House.

Norell, M. A., and P. J. Makovicky. 2001. Three cases of soft-tissue preservation in theropod dinosaurs: Changing our perspective of theropod appearance (abstr.). *J. Vert. Paleontol.* 21(3): 83A.

Norell, M., P. Makovicky, and P. Currie. 2001. The beaks of ostrich dinosaurs. *Nature* 412: 873–874.

Norell, M., J. Qiang, G. Keqin, Y. Chongxi, Z. Yibin, and W. Lixia. 2002. "Modern" feathers on a nonavian dinosaur. *Nature* 416: 36.

Norman, D. 1985. *The Illustrated Encyclopedia of Dinosaurs.* New York: Crescent.

Orr, W. N. 1986. A Norian (Late Triassic) ichthyosaur fauna from the Martin Bridge Limestone, Wallowa Mountains, Oregon. *USGS Prof. Pap.* 1435: 41–52.

Orr, W. N., and K. T. Katsura. 1985. Oregon's oldest vertebrates (Ichthyosauria [Reptilia]). *Oregon Geol.* 47(7): 75–77.

Osmolska, H. 1997. Ornithomimosauria. In *The Encyclopedia of Dinosaurs,* ed. P. J. Currie and K. Padian, 499–503. New York: Academic Press.

Osmolska, H., and R. Barsbold. 1990. Troodontidae. In *The Dinosauria,* ed. D. B. Weishampel, P. Dodson, and H. Osmolska, 259–268. Berkeley: University of California Press.

Osmolska, H., E. Roniewicz, and R. Barsbold. 1972. A new dinosaur, *Gallimimus bullatus* n. gen., sp. (Ornithomimidae), from the Upper Cretaceous of Mongolia. *Palaeontol. Polonica* 27: 103–143.

Ostrom, J. H. 1964. A reconsideration of the paleoecology of hadrosaurian dinosaurs. *Am. J. Sci.* 262: 975–997.

———. 1966. Functional morphology and evolution of the ceratopsian dinosaurs. *Evolution* 20: 290–308.

———. 1969. Osteology of *Deinonychus antirrhopus,* an unusual theropod from the Lower Cretaceous of Montana. *Peabody Mus. Nat. Hist. Bull.* 30: 1–165.

Padian, K. 1984. A large pterodactyloid pterosaur from the Two Medicine Formation (Campanian) of Montana. *J. Vert. Paleontol.* 4: 516–524.

Page, M., and R. Midgley, eds. 1993. *The Visual Dictionary of Dinosaurs.* London: Dorling Kindersley.

Parham, J. F., and T. A. Stidham. 1999. Late Cretaceous sea turtles from the Chico Formation of California. *PaleoBios* 19(3): 1–7.

Paul, G. S. 1987. The science and art of restoring the life appearance of dinosaurs and their relatives.

In *Dinosaurs Past and Present,* ed. S. J. Czerkas and E. C. Olson, 2:42. Los Angeles: Natural History Museum of Los Angeles County; Seattle: University of Washington Press.

———. 1988. *Predatory Dinosaurs of the World.* New York: Simon and Schuster.

———. 1993. [Skeletal illustration of *Euoplocephalus.*] In *The Dinosaur Society's Dinosaur Encyclopedia,* ed. D. Lessem and D. F. Glut, 189. New York: Random House.

Payne, M. B. 1951. Type Moreno Formation and overlying Eocene strata on the west side of the San Joaquin Valley, Fresno and Merced Counties, Calif. *Div. Mines, Spec. Rept.* 9: 1–29.

Plummer, F. B. 1931. Memorial of James Perrin Smith. *J. Paleontol.* 5: 168–170.

Reynolds, R. E. 1983. Jurassic trackways in the Mescal Range, San Bernardino County, California. In *Evolution of Early Mesozoic Tectonistratigraphic Environments, Southwestern Colorado Plateau to Southern Inyo Mountains,* ed. J. E. Marsolf and G. C. Dunne, 46–48. Field Guide, Geol. Soc. Am.

———. 1989. Dinosaur trackways in the lower Jurassic Aztec Sandstone of California. In *Dinosaur Tracks and Traces,* ed. D. D. Gillette and M. G. Lockley, 285–292. New York: Cambridge University Press.

Rich, T. H., R. A. Gangloff, and W. R. Hammer. 1997. Polar dinosaurs. In *The Encyclopedia of Dinosaurs,* ed. P. J. Currie and K. Padian, 562–573. New York: Academic Press.

Rieppel, O. 1999. Turtle origins. *Science* 283: 945–946.

Rieppel, O., J. Liu, and H. Bucher. 2000. The first record of a thalattosaur reptile from the Late Triassic of southern China (Guizhou Province, PR China). *J. Vert. Paleontol.* 20(3): 507–514.

Robertson, J. A. 1975. The locomotion of plesiosaurs. *Neues Jahrbuch, Geol. Palaeont. Abh.* 149(3): 286–332.

Rodríguez-de la Rosa, R. A., and F. J. Aranda-Manteca. 1999. Theropod teeth from the Late Cretaceous El Gallo Formation, Baja California, Mexico (abstract). In *VII Int. Symp. Mesozoic Terrestrial Ecosystems, Sept. 26–Oct. 1, 1999,* ed. H. A. Leanza. Buenos Aires, Arg.

———. 2000. Were there venomous theropods? (abstr.). *J. Vert. Paleontol.* 20(3): 64A.

Roth, B. 2000. Upper Cretaceous (Campanian) land snails (Gastropoda: Stylommatophora) from Washington and California. *J. Molluscan Stud.* 66: 373–381.

Rowe, T., A. B. Christopher, and K. Kyoko, eds. 1999. Cranial morphology of *Alligator mississippiensis* and phylogeny of Alligatoroidea. *J. Vert. Paleontol.,* supp. to 19(2): 9–100.

Russell, D. A. 1967. *Systematics and Morphology of American Mosasaurs (Reptilia, Sauria).* Peabody Mus. Nat. Hist. Yale Univ. Bull. 23.

———. 1970. Tyrannosaurs from the Late Cretaceous of western Canada. *Natl. Mus. Nat. Sci. Publ. Palaeontol.* 1: 1–34.

———. 1972. Ostrich dinosaurs from the Late Cretaceous of western Canada. *Can. J. Earth Sci.* 9: 375–402.

Saleeby, J. B., R. C. Speed, and M. C. Blake. 1994. Tectonic evolution of the central U.S. cordillera: A synthesis of the C1 and C2 continent-ocean transects. In *Phanerozoic Evolution of North American Continent-Ocean Transitions,* ed. R. C. Speed, 315–356. Boulder, Colo.: Geological Society of America.

Sarjeant, W. A. S. 1980. *Geologists and the History of Geology: An International Bibliography from the Origins to 1978.* New York: Arno Press.

Sarna-Wojcicki, A. M. 1995. Age, aerial extent, and paleoclimatic effects of "Lake Clyde," a mid-Pleistocene lake that formed the Corcoran Clay, Great Valley, California. In *Glacial History of the Sierra Nevada, California: A Symposium in Memorial to Clyde Wahrhaftig.*

Schile, C. A., and Abbott, P. L. 1975. Depositional environment of the Upper Cretaceous dinosaur-bearing beds along the Pacific coast west of El Rosario, Baja California. *Geol. Soc. Am., Abstracts with Programs* 7(3): 370.

Schmidt, K. P. 1938. New crocodilians from the Upper Paleocene of western Colorado. *Geol. Ser., Field Mus. Nat. Hist.* 6(21): 315–321.

Sheldon, A. M. 1996. Histological characteristics in prenatal specimens of the mosasaur *Plioplatecarpus primaevus* (abstr.). *J. Vert. Paleontol.* 16(3): 64A.

———. 1997. Ecological implications of mosasaur bone microstructure. In *Ancient Marine Reptiles,* ed. J. M. Callaway and E. L. Nicholls, 333–354. New York: Academic Press.

Simpson, G. G. Biographical memoir of Chester Stock, 1892–1950. *Natl. Acad. Sci., Biog. Memoirs,* 27: 335–362.

Smith, J. P. 1894. The metamorphic series of Shasta County, California. *J. Geol.* 2: 606.

Staebler, A. E. 1981. Survey of the fossil resources in the Panoche Hills and Ciervo Hills of western Fresno County, California. Final Rept. to U.S. Bur. Land Mgmt. in fulfillment of contract No. CA-040-CTO-10.

Staebler, C. N. 1981. A description and observations on the pelvic paddle of *Plotosaurus tuckeri.* Calif. State Univ., Fresno. Typescript.

Staebler, C. N., and A. E. Staebler. Paleontological investigations on Bureau of Land Management properties by Chad N. and Arthur E. Staebler in 1986 and 1987. Biol. Dept., Calif. State Univ., Fresno.

Stein, B. R. 2001. *On Her Own Terms: Annie Alexander and the Rise of Science in the American West.* Berkeley: University of California Press.

Stirton, R. 1955. The role of paleontology in the University of California: Honoring the twenty-fifth presidential year of President Robert Gordon Sproul. University of California, Berkeley.

Stock, C. 1938. John Campbell Merriam as scientist and philosopher. In *Cooperation in Research,* 765–778. Washington, D.C.: Carnegie Institution.

———. 1939. Occurrence of Cretaceous reptiles in the Moreno shales of the southern Coast Ranges, California. *Proc. Natl. Acad. Sci., Phil.* 25(12): 617–620.

———. 1941a. Duckbill dinosaur from the Moreno Cretaceous, California. *Abstr. Geol. Soc. Am. Bull.* 52(12): 1956.

———. 1941b. The Cretaceous vertebrate record of California. *Abstr. AAPG Bull.* 25(11): 2094.

———. 1941c. Ancient sea lizards of California. *Westways.* Jan., 21–22.

———. 1942. Sea serpents of the Panoche Hills. *Westways,* 8–9.

———. 1943. Newly discovered plesiosaur. *Eng. and Sci. Monthly* 6(11): 18.

———. 1945. John Campbell Merriam (1869–1945). *Soc. Vert. Paleontol. News Bull.* 16: 22–23, port.

———. 1946. Memorial to John Campbell Merriam (1869–1945). *Proc. Geol. Soc. Am.,* 183–198.

———. 1950. Eustace L. Furlong (1874–1950). *Soc. Vert. Paleontol. News Bull.* 29: 26–27.

———. 1951. John Campbell Merriam (1869–1945). *Proc. Geol. Soc. Am.,* 183–198, port. *Nat. Acad. Sci., Biog. Memoirs* 26: 209–217, port.

Stokstad, E. 2000. Fossils come to light in Mexico. *Science* 290: 1675.

Sues, H-D. 1997. Hypsilophodontidae. In *The Encyclopedia of Dinosaurs,* ed. P. J. Currie and K. Padian, 358–360. New York: Academic Press.

Sues, H-D., and D. B. Norman. 1990. Hypsilophodontidae, *Tenontosaurus,* and Dryosauridae. In *The Dinosauria,* ed. D. B. Weishampel, P. Dobson, and H. Osmolska, 498–509. Berkeley: University of California Press.

Sukhanov, V. B., and P. Narmandakh. 1977. The shell and limbs of *Basilemys orientalis* (Chelonia, Dermatemydae): A contribution to the morphology and evolution of the genus. *Fauna, flora i biostratgrafiya Mezozoya i Kainozoya Mongolii. Sovmestnaya sovetsko-mongol'skaya nauchneissledovat el'skaya geologicheskaya ekspeditsiya. Trudy* 4: 57–79.

Taylor, M. A. 1981. Plesiosaurs—rigging and ballasting. *Nature* 290: 628–629.

Tuminas, A. C. 1983. Structural and stratigraphic relations in the Grass Valley–Colfax area of the northern Sierra Nevada foothills, California. Ph.D. diss., Univ. Calif., Davis.

Tuminas, A. C., and E. M. Moores. 1982. Sedimentology and possible paleotectonic setting of a Late Jurassic flysch sequence in the western foothills of the northern Sierra Nevada, California. *Geol. Soc. Am. Abstracts with Programs* 14: 241.

Varricchio, D. J. 1997. Troodontidae. In *The Encyclopedia of Dinosaurs,* ed. P. J. Currie and K. Padian, 749–754. New York: Academic Press.

———. 2000. Physiological implications of reproductive behavior in the dinosaur *Troodon formosus* (abstr.). *J. Vert. Paleontol.* 20(3): 75A.

———. 2001. "Beautiful wounding tooth": Ontogeny and osteology in the theropod *Troodon formosus. Abstr. J. Vert. Paleontol.* 21(3).

Walker, W. E., Jr. 1973. The locomotor apparatus of Testudines. In *Biology of Reptilia,* vol. 4, ed. C. Gans and T. S. Parsons, 1–100. New York: Academic Press.

Wallace, J. 1993. *Familiar Dinosaurs.* The Audubon Society Pocket Guides. New York: Alfred A. Knopf; Toronto: Random House.

Wallace, R. E. 1999. *Connections: The EERI Oral History Series, Robert E. Wallace.* Interviewed by Stanley Scott. Oakland: Earthquake Engineering Research Institute.

Weishampel, D. B. 1981a. Acoustic analyses of potential vocalizations in lambeosaurine dinosaurs (Reptilia: Ornithischia): Comparative anatomy and homologies. *Paleobiol.* 7: 252–261.

———. 1981b. The nasal cavity of lambeosaurine hadrosaurids (Reptilia: Ornithischia): Comparative anatomy and homologies. *J. Paleontol.* 55: 1046–1057.

Weishampel, D. B., P. Dodson, and H. Osmolska, eds. 1990. *The Dinosauria.* Berkeley: University of California Press.

Weishampel, D. B., and Young, L. 1996. *Dinosaurs of the East Coast.* Baltimore: Johns Hopkins University Press.

Welles, S. P. 1939. Plesiosaur from the Upper Cretaceous of the San Joaquin Valley. *Abstr. Geol. Soc. Am. Bull.* 50: 1974.

———. 1943. Elasmosaurid plesiosaurs, with description of new material from California and Colorado. *Mem. Univ. Calif.* 13: 125–254.

———. 1952. A review of North American Cretaceous elasmosaurs. *Univ. Calif. Publ. in Geol. Sci.* 29: 47–144.

———. 1953. Jurassic plesiosaur vertebrae from California. *J. Paleontol.* 27: 743–744.

———. N.d. The first vertebrate from the Jurassic of the Sierra Nevada foothills, a new plesiosaur (Cryptoclididae, n.g., n.sp.). Typescript.

Wellnhofer, P. 1991. *An Illustrated Encyclopedia of Pterosaurs.* New York: Crescent.

Wieland, G. R. 1906. The osteology of Protostega. *Mem. Carnegie Mus.* 2(7): 279–304.

Williston, S. W. 1893. Mosasaurs. Part II: Restoration of clidastes. *Kansas Univ. Quarterly* 2(2): 83–84.

———. 1902. Restoration of *Dolichorhynchops osborni,* a new Cretaceous plesiosaur. *Kansas Univ. Sci. Bull.* 1(9): 241–244.

———. 1903. *North American Plesiosaurs. Part I.* Field Columbian Mus. Publ. 73. Geol. Ser. 2(1): 221–236.

———. 1925. *The Osteology of Reptiles.* Cambridge, Mass.: Harvard University Press.

Xu, X., Z. Tang, and X. Wang. 1999. A therizinosauroid dinosaur with integumentary structures from China. *Nature* 399: 350–354.

Xu, X., X. Wang, and X. Wu. 1999. A dromaeosaurid dinosaur with a filamentous integument from the Yixian Formation of China. *Nature* 401: 262–266.

Xu, X., Z. Zhou, and X. Wang. 2000. The smallest non-avian theropod dinosaur. *Nature* 408: 705–707.

Yang, D. 1981. A large mosasaur from the Cretaceous deposits of the Cantua Creek area of California. Calif. State Univ., Fresno. (Unpub., lost.)

———. 1983. A study of the pectoral and pelvic appendages of California mosasaurs. Master's thesis, Calif. State Univ., Fresno.

Zangerl, R. 1953. *The Vertebrate Fauna of the Selma Formation of Alabama.* Part 4: *The Turtles of the Family Toxochelyidae.* Fieldian Geol. Mem. 3(3,4). Chicago Nat. Hist. Mus.

Zullo, J. L. 1969. Annie Montague Alexander: Her work in paleontology. *J. of the West* 8(2): 183–199.

INDEX

Numbers in italics refer to pages with illustrations.

California Academy of Sciences, 152, 171, 188, 246; *Hydrotherosaurus alexandrae*, 103, *104*

California, Alta, 230, 232, 243

California Geologic Survey, 167

California Institute of Technology. *See* Caltech

California State College, San Diego, 253

California State University, Chico, 149, 150, 156, 210

California State University, Fresno, 103, 108, 171, 179, 213, 214, 216, 220, *219*

California State University, Los Angeles, 241

California, Upper. *See* Alta California

Californosaurus, 88, 90, 93, *93*

Californosaurus perrini, 84, 93, *30*, *93, 142*

Calkin, Martin, 170

Caltech, 179, 197, 210, 211; Buwalda at, 135; Drescher at, 182, 183; Furlong, E. at, 102, 129, 135, 139, 207; laboratory, *206;* Leard at, 200; and *Morenosaurus stocki*, 105, Mudd Geology Building, 208; *Plesiotylosaurus crassidens* in lab, *185;* and *Plotosaurus tuckeri*, 193, *195*; Stock at, 135, 139, 168, 173, 178, 188, 189

Calvano, Gino, 114, 241, 242, *242*

Camp, Charles L., 173, 179, 188, 189, 206, 207, *174;* and *California Mosasaurs*, 170

Camp, Harriet, 173

Campbell, Arthur S., 94

carapace: *Basilemys*, 114, 115, *120;* Chelonoididea, 118; dermochelyid turtle, 116, *120;* leatherback, 115, 116; *Osteopygis*, 119, *120;* toxochelyid, *120;* turtle, 164, 239, *164*

carcass(es), 21, 26; ankylosaur, 40, 238; hadrosaur, 243; ichthyosaur(s), 225; plesiosaur(s), 172, 225

Carlsbad, 39, 40, 235, 236, 241

Carnegie Institute in Washington, D.C., 135, 139, 182, 189

carnivorous dinosaur(s): *Albertosaurus*, 53, 54; *Anchisauripus*, 229; dromaeosaurid, 54; *Gorgosaurus*, 248; *Grallator*, 229; *Labocania anomala*, 53, 54, 254; ornithomimid, 54; *Saurornitholestes*, 53, 54; theropod(s), 31, 35, 37, 44, 53, 254; *Troödon formosus*, 53, 54; tyrannosaurid, 54

carnosaur(s), 56, 250, 251, *56, 57*

carnosauria, *30*

Carquinez Strait, 131

Carranza, Oscar, 254

Carter, Raymond, 143

Cascade/Modoc Province, 125, 127, 149

Case, Gerard R., 152, *152*

Case, Leon, 234

Caudipteryx, 74

caves, 233. *See also* limestone(s)

Central Valley, 166

Centrosaurus, 43

cephalopod(s), 87. *See also* ammonite; belemnite; chambered nautilus; octopus; squid

Ceraopoda, *30*

ceratopsian, 42, 43, *43*

Cerutti, Richard, 235–37, *235*

chambered nautilus, 81, 82, 97

Chandler, Robert, *235*

Chaohusaurus, 84

Chelonia, 154

Cheloniidae, 115, 117–19, 155, *30*

chelonioid, 115, 117, 118, 157, *30*

Chelonioidea, 115, *30*

Chelonoididea, 115

chert, 221. *See also* radiolarian chert

Chico Formation, 156, 157; bird(s), 75, 155; Chelonia, 154; Cheloniidae, 118, 155; *Clidastes*, 111, 155, 157; dermochelyid, 116, 155; fern(s), 26, 74; fossil evidence, 21, 26; *Ichthyornis* humerus, *158;* leaves, 71, 157; plesiosaur tooth, 97; pterosaur wing bone, *158;* pterosaur(s), 26, 71; redwood(s), 26, 74; reptile tooth, 149, 150; shark teeth, 71, 157; shells, 71, 157; theropod, 155; Toxochelyidae,

119; tree fern, *54;* trees, flowering, 26, 74; turtle carapace, *164;* turtle(s), 155

China, 80, 88; Lianoning Province, 74

Chinle Formation, Arizona, 213

Christe, Geoff, 38, 147, 149, *38*

Cincinnati Museum Center, 98, 166

Claque, John, 248

Clark, Bruce L., 141–43

Clark, Lorin D., 146, 147, *147*

Clark Park in Buena Park, 241

Clark, Richard, 248, *250*

claw(s): *Albertosaurus*, 54; bird(s), 74; dew, 31; dinosaur, 31; dromaeosaurid, 59; mosasaur, 220; ornithomimid, 62; *Thalattosaurus alexandrae*, 80, *81; Troödon*, *57;* wing (bird), 74

Clidastes, 111, 157, *112, 113*

"Clipper Gap Monster," 145, *146*

club, tail, 41, 42, *40, 42*

Coalinga earthquake, 218, 220, *220*

coastal: desert, 18, 37, 229, 244; dune(s) in Africa, 25; fluvial deposits, 45; redwood forest, 25

Coast Ranges Province, 149, 223–25; age dating, 24; Franciscan Formation, 22–24, 94, 98, 221; ichthyosaur(s), 94, 95, 168, 224, 225; plesiosaur(s), 225; rain shadow, 167; reptilian remains, 224

Coast Range Thrust (fault), 127

Coelurosauria, *30*

Coker, Tom, *91*

Colbert, E. H., 246

Colman, Royce, 227

Coelurosauria, 40

concretion(s): hadrosaur vertebra, 233, 236; hypsilophodontid, 51, 154; limestone, 146, 225, 234; mosasaur vertebrae, 234; plesiosaur, 146; siltstone, 237

cone-bearing. *See* conifer(s)

conglomerate(s), 13; Baja, 243; Cabrillo Formation, 239; Late Jurassic, 21, *18;* pebbly, 166; Trail Formation, 25; Triassic, 24

California, 37, 229; egg(s), 44, 51, 58; eggshell, 254; *Euoplocephalus*, 41, 42, *30, 42*; evolution of, 5, 29, 31, 57; exploration, 227; extinction of, 7, 222; family tree, *30;* feathers, 35, 60, 61, *59, 60;* feeding habits, 40, 42; finds by county, *36;* first in Baja, *247;* first in world, 29; first in California, 170, *2;* first discovery site, *170, 171;* foot (feet), 31, 57, 241, *32;* footprints, 149; *Gallimimus,* 62; *Grallator,* 37, *37;* gregarious nature of, 40; habitat, 36, 53; hand(s), 31, 35, 54, 57, *32;* hearing, 58; hips, 34, *35, 252; Hypacrosaurus altispinus,* 48; *Hypsilophodon,* 51; ichnogenera, 37; ischium, 34; jaw, 35, 55, 56, 58, *218;* Jurassic, 36–38; Jurassic habitat, 36; Jurassic rib, *38; Kritosaurus,* 47; *Labocania anomala,* 53, 54, 55, 56, *57;* Lambeosaur, *231; Lambeosaurus laticaudus,* 43–49, 51, *50; Lambeosaurus laticaudus* ischium, *252; Lambeosaurus laticaudus* spine, *48; Lambeosaurus* skin mold, *244; Lambeosaurus laticaudus* skull, *48;* leg(s), 31; lizard-hipped, 34, 35, *35;* locomotion, 31, *32; Maiasaura,* 55; meat-eating, 26, 27, 35, 155; metapodial, 239, 241; metatarsal 37, 38, 162, *163;* models, scale, 208; Mojave Desert, 36, 37, *37; Monoclonius, 43;* mummified remains, 34; nest(s), 44; news article, *2;* Ornithischia, 34; ornithomimid, 54, 61–63, *63;* ornithomimid skeleton, *62;* ornithomimid skull, *63;* ostrich, *63; Pachycephalosaurus, 33; Parasaurolophus, 44; Psittacosaurus,* 42; as prey, 44, 49, 51; pubis, 34; quadruped, 31; reconstruction(s), 33; relationship to birds, 35; remains, 25, 242, 243; San Diego County, 26; Saurischia, 34; sauropod,

217; *Saurornitholestes,* 53, 54, 60, 61, *60; Saurornitholestes* feathers, *60;* sexual display, 42, 46, 48; Sierra Nevada, 38, 147, 149; *Sinornithosaurus millenii,* 60; skeletal remains, 19, 247; skin impressions, 34, 43, 243, *244; Stegasaurus* fin, 217; tail tendons, 45; teeth, 35, 38–40, 42, 43, 46, 51, 55, 60, 149, 246, 248, 250, *44, 233, 247; Thescelosaurus,* 51; tibia, 250, 251, *250, 251;* toe configuration, 31; tracks (trackways), 25, 36, 37, 227, 229, *37, 228, 230;* taxonomy, 34, 35; Triassic, 36; *Triceratops,* 42, 43; *Troödon,* 57, 58; *Troödon formosus,* 53, 54, 57, 58; Troödontid, *58, 59;* Tyrannosaurid, 54; *Tyrannosaurus rex,* 57; *Tyrannosaurus rex* skull, *34;* vertebra, 149, *250; Velociraptor mongoliensis,* 60; vestigial digits, 31; vision, 55, 58, 61; vocalization, 46, 48; word origin, 29. *See also Albertosaurus;* ankylosaur(id); carnivorous dinosaurs; hadrosaur; herbivorous dinosaur; hypsilophodontid; *Saurolophus;* skeleton; skull; theropod

Dinosaur Point, 216, 217, *216*
Dinosauria, *30*
Dinosauriformes, *30*
Dinosauromorpha, *30*
dog. *See* Temy
Dolichorhynchops, 98, 151, *30, 99*
Dolichorhynchops osborni, 98
Dos Palos, 180
Dougherty, Jack F., 179, 182, 185, *183*
Downing (Desatoff), Marsha, 215
Doyle, Peter, 85
Drachuk, Robert, 234
Drescher, Arthur (Art), 189, 200, 211; *Aphrosaurus furlongi* find, *183;* biography, 182, 183; in Fresno, *180, 182; Morenosaurus stocki,* 105, *184, 199;* in Panoche Hills, 178–82, 184, 185, 186, 194, 195, 197, 201, 202, 204–6; and

Plesiotylosaurus crassidens, 182, 185; and *Plotosaurus tuckeri, 190, 191, 201;* at Reptile Ridge Camp, *186;* and *Saurolophus,* 201, 202, 204, 205
dromaeosaurid, 61, *32*
dromeosaurid, 54, 59–61, 74, *30*
dromaeosauridae, *30*
duck-billed dinosaur. *See* hadrosaur
dune, 229; in Africa, 25; in Baja, 250; coastal, 25; cross-bedded sand, 18, 37, 229; Jurassic dinosaur tracks, 25, *228, 230*
Durham, J. Wyatt, 47, 98, 225, 246

earthquake(s), 210, 218, 220, 141, 188, *220*
East Park Reservoir, *150*
Eastern Klamath Mountains Province, 13
Eastern Transportation Corridor, 114, 241
egg(s): botflies, 243; hadrosaur, 44, 51; *Troödon,* 58; shells, 254
El Gallo Formation, 41; ceratopsian, 42, 43; crocodilian(s) in, 64, 65; dinosaur(s) in, 246; hadrosaur in, 47; theropod in, 55, 63, 64
El Rosario, Baja, 248, 251–53
elasmosaur(id)(s), 20, 98, 101, 221, 227, 233
Elk Creek (creek), 152, 164
Elk Creek (town), 97, 98, 150, 152, 161, 164
Embree, Pat, 71, 145, 154, 155, 157, 158, *154*
Enantiornithes, *30*
England, first dinosaur fossils in, 29
environment(s), 24–27
environmental impact report(s), 234, 235
Eoichthyosauria, *30*
Eosauropterygia, *30*
Esquela Superior de Ciencias Marinas, 247, 248
Erickson, Gregory, 155
Ervin, Steven, 220, *218, 220*
Esterly, W. B., 140

mapping, 164, 171, 194, 215

Marin Headlands, 23, *23*

marine: depositional settings, 20–24, 26–27, 243; rock(s), 51, 227, 232, 242

marine invertebrates. *See* invertebrates(s)

marine reptiles, 79–121; behavior of, 79; crocodiles, *66;* early, 20; evolution of, 79; extinction of, 222; ichthyosaur(s), 84–95; iguana, *79;* mosasaur(s), 106–12; plesiosaur(s), 97–195; thalattosaur(s), 80–83; turtle(s), 113–21; *117, 118, 120. See also individual marine reptile names*

Mariposa County, plesiosaur, 101, 146

Mariposa Formation, 20, 21, 101, 146

Matson Shipping Lines, 129, 134

Matteson Ranch, 130, 132, 133,

Matthew, William D., 178

Mattison, Dave, 159, 163

maxilla, hadrosaur, 233, *152, 233*

May, Clifford, 160

McCaskill, Letty (Betty Lee) E., 201

McClure Reservoir, 146

McDougall, Charles, 179, 203, 204

McKee, Bates, 98, 151

McKensie, Theodocia Wilmoth, 171

meat-eating. *See* carnivorous dinosaur(s)

Mendota, 180, 204, 205, 208

Merced County, 107, 216

mercury mines, 167

Mercy Hot Springs, 180

Merriam, John C., 102, 145, 188, 189, 206, 207, *138;* biography, 138, 139; and *Californosaurus* skeletons, 93; and cohorts, 129–43; and Euichthyosauria taxonomy, 91–93; expeditions of, 129, 131, 134; at Rancho La Brea, 139; *Shastasaurus pacificus* described by, 88; and thalattosaur taxonomy, 80; tooth, described by, 149, 150; *Triassic*

Ichthyosauria with Special Reference to the American Forms, 142; and vertebrae, 150

Merriamia zitteli, 91–93, *92*

Merriamosauria, 139, *30*

Mesodermochelys, 117

Mesozoic, 5; Atlantic Ocean, 10; California, 9, *15;* evolution, 7; fossil(s), 25; K-T catastrophe, 7; plate tectonics, 12, *16;* reptile family tree, *30;* reptile fossil preservation, 17; reptile tracks, 25; terrain, 24; terrestrial environments, 24–27

metacarpal, pterosaur, 71–73, 157, 158, *72*

metapodial, 239

metatarsal, dinosaur, 37–38, 51, 162, *163*

meteorite, 222

Metornnithes, 30

Mexico, 88, 200, 230, 243, 246

Michael, Joseph, 172

microfossils, 20, 23, *24*

Mill Creek Canyon, 154

Miller, Hugh, 130

Miller, Loye H., 143

mines, mercury, 167

Misty Hill, 244

Mitchell, Ed, 213

Mixosaurus, 90

Modesto, 212

Modesto Junior College, 212, *212*

Moise, Woodrow (Woody), 215, 217

Mojave Desert Province, 138, 227, 229, *228;* dinosaur tracks in, 36, 243, *37, 228, 230;* Jurassic, 18, 25

Molina, Adolfo, 248

mollusk, 7, 71, 75, 106, 153, 156, 235, 238

monitoring, paleontological, 235, 236, 241, 242

monkey puzzle tree *(Araucaria),* 26, 27, 53, 244

Monoclonius, 43

Mononykus, hand(s), 31

Montana, 49, 54, 57, 157, 217, 245

Monterey Peninsula College, 212, *211*

Moore, B., 233, *233*

Moore, E. W., *173, 193*

Moreno Formation, 166; *Aphrosaurus furlongi,* 102; *Basilemys,* 114; elasmosaur, 221; *Fresnosaurus drescheri,* 102; hadrosaur(s), 46, 47; *Hydrotherosaurus alexandrae,* 103; mapping of, 215; *Morenosaurus stocki,* 105; *Osteopygis,* 119; plesiosaur, 102; *Plesiotylosaurus crassidens,* 110; plotosaur, 168; *Plotosaurus bennisoni,* 107; pterosaur, 71, 156, 157; stratigraphic relationships of, 215

Morenosaurus, 102

Morenosaurus stocki, 105, 182, 200, 207, 212, 213, *105;* at Caltech, 105; characteristics of, 105; excavation of, *184, 198, 199;* family tree, *30;* flipper, 197; Fresno County, 105; gastroliths, 105; girdle, 197; in Moreno Formation, 105; namesake, 105, 189; on sled, *202;* in Panoche Hills, 105; preparation, *212;* quarry, *198,* skeleton, 105, 184, 208, *184, 209;* skull, 207, 208, *209;* tail, 197; teeth, 212, 213; vertebrae, *184*

Morris, William J., *252;* in Baja, 245–54; biography, 252, 253; crocodilian find, 66; dinosaur find(s), 247, 249, *248, 250;* hadrosaur find, 47, 48

mosasaur(s), 106–12, 138, 157, 179, 189, 211, 215–17, *165;* "baby" *Plotosaurus,* 212, *211;* behavior of, 106; characteristics of, 106; in Chico Formation, 155; claw, 220; evolution of, 87, 106; excavation of, *181, 214, 218;* extinction of, 106; family tree, *30;* finds by county, 106, *107;* flipper(s), 215, 218, 220; in Granite Bay, 155; hearing, 106; jaw(s), 106; in Panoche Hills, 182, 183, 200, 212, 223; preparation, *219;* prey, 106, 153; remains, 213, 215, 218, 220, 234; reproduction, 106–7; scapula, 206; skeletal remains, 19, 106,

Point Loma, 235, 240
Point Loma Formation, 39, 233–36, 239
Poppadakis, Steven, 254
Pray, Lloyd C., 189
preparator(s), 124; Alexander, Annie Montague, 135; Calkin, Martin, 170; Cerutti, Richard, 235, 236; Drescher, Arthur, 182; Embree, Pat, *154;* Furlong, Eustace, 206, 207, 208, 131, 141, 142, *209;* Garbani, James (Harley), 247, 248; Howard, Don, 212; Maloney, David, 156, 160; Matthew, William D., 178; Miller, Loye H., 143; Otto, William, 105, 207, *195, 209;* Riney, Bradford, 236, 241; Rope, Mr. 143; Takeuchi, Gary, 241
Princeton University, 131, 152, 252
propodial, 147
Protarchaeopteryx, 74
provinces, geologic, 124, *124;* northern, 127–25; southern, 227–56. *See also individual province names*
Pseudosuchia, *30*
Psittacosaurus, 42
Pteranodon, 71–73, *70, 73, 144*
Pteranodon ingens, 73
Pteranodon sternbergi, 72, 73
Pterosauria, *30*
pterosaur(s), 26, 69–74, 157, *144;* azhdarchid, 71; behavior of, 69, 71; in Budden Canyon Formation, 71; in Butte County, 71–74, 164; in Chico Formation, 26, 71; Cretaceous, 71, 72; evolution of, 6; extinction of, 222; family tree, *30;* flight, 69; in Great Valley Group, 71; habitat, 71, 74; humerus, 71; in Klamath Mountains, 71; leg bone(s), 71; metacarpal, 71–73, 157, 158, *72;* in Moreno Formation, 71, 156, 157; nesting, 71; in Oregon, 71; in Panoche Hills, 71, 156, 157; *Pteranodon,* 71–73, *73; Quetzalcoatlus,* 71, 72; in San Joaquin Valley, 71, 156, 157; in Shasta County,

71, 158; size, 71–73; in South America, 69, 71; ulna, 71, 73, *72;* vertebrae, 71; wing bone(s), 69, 71, 156–58; *72, 158*
pubis, dinosaur, 34, 35, 59
pustulse turtle. See *Naomichelys*

quadruped, 31, 39
Queenie (mule), 185, *187*
Quetzalcoatlus, 71, 72

radiolaria(n) , 23, 24, 94, 224 *24*
radiolarian chert, 22–24, 94, 225; acid etching, 24; cobble, 22, 94, 95, 168, 207; definition of, 224; formation, 94; in Franciscan Formation, 22, 94; ichthyosaur(s), 94, 95, *224;* iron, 23; on Marin Headlands 23, *23;* plate tectonic effects, 94; reptilian remains, 224; rust, 23; sediment, 23, 94; silicon dioxide, 23
rain shadow desert, 229
Ralph B. Clark Park in Buena Park, 241
Rancho La Brea, 139, 188
rattlesnake(s), 134, 159
Rausch, Margo, 237
Ray, F. S., 140
Ray, Mrs. J., 210
Redding, 51
redwood, 25–27, 244
reef(s), 13, 40, 238
reefal limestones. *See* limestone(s)
regurgitation, ichthyosaur, 85, 86
rip-up clasts, 21
Reptile Ridge Camp, 185, *186*
reptile(s), aquatic: Baja, 27; evolution of, 7, 29; family tree, *30;* flying, *70, 73;* fossil preservation, 17; marine, 20, 166; Permian, *33;* skeleton, 13; tracks, 25
reptilian fossils (remains), 25; in Baja, 246, 254; in Coast Range, 224, 225; Cretaceous, 19, 22; in flysch, 21; in Great Valley Group, 2; Jurassic, 19, 22. *See also individual reptile names*
Reynolds, Robert E. (Bob), 37, 229, *230*

ridge(s), 9, 10, 11, 230, *10, 11*
rifting, 14, 17, 230, *16*
Riney, Bradford (Brad), *240;* ankylosaur find, 39, 236, 237, *238;* biography, 240, 241; dinosaur foot bone find, 236; hadrosaur finds, 233, 235, 236, *235, 236;* mosasaur finds, 23, 236
river(s), 26; Baja, Cretaceous, 243; Cretaceous, 19, 71; deposits in, 48, 49, 149; Triassic, 25
Rocklin, 155
rocks: Cretaceous sedimentary, 26; Franciscan Formation, 22; Late Jurassic, 25, 149; marine, 51, 227, 232; metamorphic, 71, 145, 149; mudstone, 236; oldest Jurassic marine, 20; sedimentary, 20, 21, 26, 244; siltstone, 19, 154, 237; terrestrial, 38; volcanic, 149. *See also* limestone(s); sandstone(s)
Rodda, Peter, 152, 153, *153*
Roeder, Mark, 234, 236, 241, 242, *242*
Rogers, Tom, 146
Rominger, Joe, 179, 185, 195 *186, 192*
Roper, Norvel, 179, 201
rostrum, ichthyosaur, 86, 90, 94, 95, 207, 224, 225, *95, 224*
Royal Tyrrell Museum of Paleontology, Alberta, Canada, 80, 162

Sacramento Bee, 151
Sacramento Junior College, 212
Sacramento Sunday News, 146
Sacramento Valley, 159; Chelonia, 154; Chico Formation, 155; flysch, 21; granite, 18; Great Valley Group, 21; hogbacks, 18, 149; Jurassic plesiosaur(s), 97, 98, 162, 225; linear hills, 21, 149; rainfall, 149; reptile(s), 151; sediments, 149; strike valley(s), 149
sacrum, *Lambeosaurus,* 249
Salt Springs Slate, 20, 146
Salton Trough, 230
Salton Province, 125

vestigial digits, 31
vine(s), 27, 244
vision: *Albertosaurus,* 55; ichthyo-
    saur, 86; mosasaur(s); 106;
    ornithomimid, 61; *Troödon,* 58
vocalization, 46, 48
von Zittel, Karl, 138

W. M. Keck Museum, 103
Wahl, Karl, 151
Wahl-Hall, Donna Rae, 151, 159
Wahrhaftig, Clyde, 179, 186–88,
    195, 197, 206, *186, 192, 196;* biog-
    raphy, 210, 211; mother of, 188
Walker, Herman G., 168
Wallace, Robert, 105, 179, 182,
    186, *184, 185, 190;* biography,
    211
Warren, Ryan, 161
Wegener, Alfred, 12
Welles, Harriet, *175, 176*

Welles, Samuel P., 135, 170, 189,
    206, 207, 208, 210–13, 215, 217,
    170, 175, 178; biography, 178,
    179; *Fresnosaurus drescheri*
    find, 182; hadrosaur find, 169;
    Hydrotherosaurus alexandrae
    find, *178; Morenosaurus stocki*
    find, 213; mosasaur find, 210,
    215, *218;* plesiosaur finds, 101,
    146, 147, 172–74, 177, 178;
    Plesiosaurus hesterus find, 98,
    101; *Plotosaurus bennisoni* skull
    find, 213, *170;* in Shasta
    County, 143
Wemple, Edna, 140
Wetmore, Alexander, 75
Whistler, David, 241
White, B. F., 143
White, Robert T., 182
Wiggins, Ira, 246
Williams Formation, 241, 242

Wilmoth, Bob (Potsey), 179,
    203
Wilson, Robert (Bob), 179, *193;*
    biography, 211
windstorm, 181
wing(s), 69, 71, 74, 156–58, *72,
    158*
Wood, Jason, 85
wood, 21, 26, 166, 250
World War II, 201, 208, 210
Wu Xiao-Chun, 162

Yale Peabody Museum, 152
Yang-Staebler, Dianne, 213, 215,
    217, 223, *218, 222;* biography,
    216, 217
Yarborough, Mark, 215
Young, G., 151
Yucatan, 222, 223

Zarconi, Carl, 213, *212*

DESIGNER	Victoria Kuskowski
COMPOSITOR	Integrated Composition Systems
TEXT	11/16 Adobe Garamond
DISPLAY	Univers Condensed Regular, Egiziano, Stymie Condensed
PRINTER AND BINDER	Imago

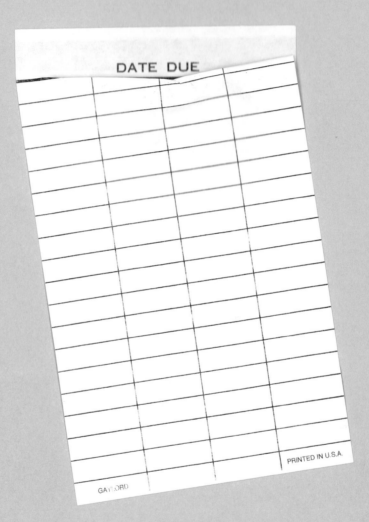

DATE DUE

PRINTED IN U.S.A.

GAYLORD